Energy, Climate and the Environment Series

Series Editor: **David Elliott**, Emeritus Professor of Technology Policy, Open University, UK

Titles include:

Manuela Achilles and Dana Elzey (*editors*)
ENVIRONMENTAL SUSTAINABILITY IN TRANSATLANTIC PERSPECTIVE
A Multidisciplinary Approach

Luca Anceschi and Jonathan Symons (*editors*)
ENERGY SECURITY IN THE ERA OF CLIMATE CHANGE
The Asia-Pacific Experience

Philip Andrews-Speed
THE GOVERNANCE OF ENERGY IN CHINA
Implications for Future Sustainability

Ian Bailey and Hugh Compston (*editors*)
FEELING THE HEAT
The Politics of Climate Policy in Rapidly Industrializing Countries

Gawdat Bahgat
ALTERNATIVE ENERGY IN THE MIDDLE EAST

Mehmet Efe Biresselioglu
EUROPEAN ENERGY SECURITY
Turkey's Future Role and Impact

David Elliott (*editor*)
NUCLEAR OR NOT?
Does Nuclear Power Have a Place in a Sustainable Future?

David Elliott (*editor*)
SUSTAINABLE ENERGY
Opportunities and Limitations

Huong Ha and Tek Nath Dhakal (*editors*)
GOVERNANCE APPROACHES TO MITIGATION OF AND ADAPTATION TO
CLIMATE CHANGE IN ASIA

Horace Herring and Steve Sorrell (*editors*)
ENERGY EFFICIENCY AND SUSTAINABLE CONSUMPTION
The Rebound Effect

Horace Herring (*editor*)
LIVING IN A LOW-CARBON SOCIETY IN 2050

Matti Kojo and Tapio Litmanen (*editors*)
THE RENEWAL OF NUCLEAR POWER IN FINLAND

Antonio Marquina (*editor*)
GLOBAL WARMING AND CLIMATE CHANGE
Prospects and Policies in Asia and Europe

Catherine Mitchell, Jim Watson and Jessica Whiting (*editors*)
NEW CHALLENGES IN ENERGY SECURITY
The UK in a Multipolar World

Catherine Mitchell, Jim Watson and Jessica Whiting (*editors*)
NEW CHALLENGES IN ENERGY SECURITY
The UK in a Multipolar World

Catherine Mitchell
THE POLITICAL ECONOMY OF SUSTAINABLE ENERGY

Ivan Scrase and Gordon MacKerron (*editors*)
ENERGY FOR THE FUTURE
A New Agenda

Gill Seyfang
SUSTAINABLE CONSUMPTION, COMMUNITY ACTION AND THE NEW
ECONOMICS
Seeds of Change

Benjamin K. Sovacool
ENERGY & ETHICS
Justice and the Global Energy Challenge

Joseph Szarka
WIND POWER IN EUROPE
Politics, Business and Society

Joseph Szarka, Richard Cowell, Geraint Ellis, Peter A. Strachan and Charles
Warren (*editors*)
LEARNING FROM WIND POWER
Governance, Societal and Policy Perspectives on Sustainable Energy

David Toke
ECOLOGICAL MODERNISATION AND RENEWABLE ENERGY

Thijs Van de Graaf
THE POLITICS AND INSTITUTIONS OF GLOBAL ENERGY GOVERNANCE

Xu Yi-chong (*editor*)
NUCLEAR ENERGY DEVELOPMENT IN ASIA
Problems and Prospects

Xu Yi-chong
THE POLITICS OF NUCLEAR ENERGY IN CHINA

Energy, Climate and the Environment
Series Standing Order ISBN 978–0–230–00800–7 (hardback)
978–0–230–22150–5 (paperback)
(*outside North America only*)

You can receive future titles in this series as they are published by placing a standing
order. Please contact your bookseller or, in case of difficulty, write to us at the address
below with your name and address, the title of the series and the ISBNs quoted above.

Customer Services Department, Macmillan Distribution Ltd, Houndmills, Basingstoke,
Hampshire RG21 6XS, England

Governance Approaches to Mitigation of and Adaptation to Climate Change in Asia

Edited by

Huong Ha

Academic Coordinator, University of Newcastle, Singapore

and

Tek Nath Dhakal

Professor, Tribhuvan University, Nepal

First published 2013 by
PALGRAVE MACMILLAN

Palgrave Macmillan in the UK is an imprint of Macmillan Publishers Limited, registered in England, company number 785998, of Houndmills, Basingstoke, Hampshire RG21 6XS.

Palgrave Macmillan in the US is a division of St Martin's Press LLC, 175 Fifth Avenue, New York, NY 10010.

Palgrave Macmillan is the global academic imprint of the above companies and has companies and representatives throughout the world.

Palgrave® and Macmillan® are registered trademarks in the United States, the United Kingdom, Europe and other countries.

ISBN 978–1–137–32520–4

This book is printed on paper suitable for recycling and made from fully managed and sustained forest sources. Logging, pulping and manufacturing processes are expected to conform to the environmental regulations of the country of origin.

A catalogue record for this book is available from the British Library.

A catalog record for this book is available from the Library of Congress.

Contents

Figures and Tables

Figures

Tables

Series Editor's Preface

Energy, Climate and the Environment

Concerns about the potential environmental, social and economic impacts of climate change (CC) have led to a major international debate over what could and should be done to reduce emissions of greenhouse gases (GHGs). There is still a scientific debate over the likely scale of CC, and the complex interactions between human activities and climate systems, but global average temperatures have risen and the cause is almost certainly the observed build-up of atmospheric GHGs.

Whatever we now do, there will have to be a lot of social and economic adaptation to CC – preparing for increased flooding and other climate-related problems. However, the more fundamental response is to try to reduce or avoid the human activities that are causing CC. That means, primarily, trying to reduce or eliminate emissions of GHGs from the combustion of fossil fuels. Given that around 80 per cent of the energy used in the world at present comes from these sources, this will be a major technological, economic and political undertaking. It will involve reducing demand for energy (via lifestyle choice changes – and policies enabling such choices to be made), producing and using whatever energy we still need more efficiently (getting more from less) and supplying the reduced amount of energy from non-fossil sources (basically switching over to renewables and/or nuclear power).

Each of these options opens up a range of social, economic and environmental issues. Industrial society and modern consumer cultures have been based on the ever-expanding use of fossil fuels, so the changes required will inevitably be challenging. Perhaps equally inevitable are disagreements and conflicts over the merits and demerits of the various options and in relation to strategies and policies for pursuing them. These conflicts and associated debates sometimes concern technical issues, but there are usually also underlying political and ideological commitments and agendas which shape, or at least colour, the ostensibly technical debates. In particular, at times, technical assertions can be used to buttress specific policy frameworks in ways which subsequently prove to be flawed.

The aim of this series is to provide texts which lay out the technical, environmental and political issues relating to the various proposed policies for responding to CC. The focus is not primarily on the science of CC, or on the technological detail, although there will be accounts of the state of the art, to aid assessment of the viability of the various options. However, the main focus is the policy conflicts over which strategy to pursue. The series adopts a critical approach and attempts to identify flaws in emerging policies, propositions and assertions. In particular, it seeks to illuminate counterintuitive assessments, conclusions and new perspectives. The aim is not simply to map the debates but to explore their structure, their underlying assumptions and their limitations. Texts are incisive and authoritative sources of critical analysis and commentary, indicating clearly the divergent views that have emerged and also identifying the shortcomings of these views. However, the books do not simply provide an overview; they also offer policy prescriptions.

The present text, which focuses on energy/climate policy and governance issues in Asia, certainly includes some recommendations, as well as a range of insights into the problems of deploying effective policies. The emerging economies of Asia are clearly faced with major political, economic and social challenges as CC begins to impact, and this overview looks at some of the key issues, policies and processes for mitigation and adaptation. The emphasis is on governance and strategic management rather than the details of specific technological, economic or social policies, although renewable energy development is clearly seen as a key element. Given the unevenly developed nature of some Asian economies, the political problems of making the changes needed are recognised, but the overall message is a positive one. Indeed, as one contribution suggests, it may be that the North, with its unsustainable consumption of non-renewable resources, may have much to learn from examples of low-carbon economies in the South.

Foreword

This is a timely book, insofar as the four-year cycle of the Intergovernmental Panel on Climate Change is to issue a new compendium of global facts and forecasts in September 2013. However it focuses on Asian matters, concentrating on the social science issues involved in decisions on CC. Of course, as the editors Huong Ha and Tek Nath Dhakal stress, mitigation and adaptation are both serious and pressing issues, yet in relation to CC they must coexist with policies that deal with the totality of a nation's issues. I think that many people in recent years have thought less about CC issues as a result of their preoccupation with financial problems following the US-led crash of 2006. This has diverted action away from the important long-term thinking required for CC mitigation and adaptation.

The CC policies derived by nations all differ. This causes friction in the international meetings as each leader proposes a different action plan as they wish to gain the best deal for their citizens – which may be a 'firewall', such as an easement in restrictive practices on emissions control. The leaders see their results as 'winning', but possibly they ought to be compromising for the good of the whole. We see such fights reported by the media at the annual Conference of all Parties (COP) meetings organised by the United Nations (UN) Framework Convention on Climate Change (UNFCCC). The UN proclaimed some meetings as 'important', like the famous one in Kyoto (1997); or the later ones that were to lead to new agreements for the renewed control of CC – like Bali (2007), Copenhagen (2009) or Doha (2012), which all overran their scheduled time and, in reality, ended in chaos with, I suggest, no real outcomes. I accept that at Doha there was a document signed off by all nations that ought to lead to a new revision of the Kyoto Protocol by 2020, but it is weakly worded.

Several of this book's authors have addressed the squabbling at these meetings. In particular, Gamini Herath in Chapter 5 draws together the developed and developing nations' difficulties and the relative failure of the Kyoto Protocol. Huong Ha argues that the mechanisms of these 'big' meetings need to be streamlined. This is a difficult process, of course, involving considerable insights which need to be developed via systems sciences as well as through intense political discussions. Kidd (2013) argues that it may be possible to make the COP meetings considerably more effective if the UN accepts that its secretariat could engage in political matters. He also argues that the complexity of the very many valid national aims calls for a redesign of the structure of the UN meetings so that the hapless host nation is not foisted with the impossibility of managing the logistics of the meetings (with over 15,000 attendees in some cases) as well as being tasked with ensuring an accord is developed among all 196 nations' discussants. This book illustrates some of the issues and the complexities brought forward from the Asian viewpoint.

Let me draw your attention to an issue located far away from Asia – the melting ice caps of the Arctic and Antarctic. Science indicates that global ice packs are melting (at the North and South Poles, and at the many glaciers). The northern Arctic melt will have little effect (as with ice cubes floating and melting in a glass of water, no overflow takes place); but the increased sunlight on newly exposed water will cause it to expand as water is most dense at 4°C. Even with the small consequent global sea-level rise, many low-lying islands will be flooded, and the people thereon will totally lose their national state. There is no forum in the UNFCCC to discuss 'displaced persons', yet in the 2011 Durban meeting many statesmen argued for more stringent pollution controls as the sea levels were rising and their countries would be flooded, effectively leaving all of their people stateless. I too opined (11 June 2012) that few seemed to care – certainly not the general public, as their concern about the Arctic and Antarctic is limited to the "nice penguins and polar bears!" and not to any melting that may be taking place. This malaise has now spread to the Second Earth Summit as Rio+20 having a '*zero* draft text' issued by the UN Conference on Sustainable Development (UNCSD) in January 2012, which suggested that leaders would not have to sign anything that was not in the original Rio declaration of 20 years ago. I don't think this is progress.

The scientists in the Antarctic find it is also melting, and as this is on-land glacial cover its freshwater will lift sea levels. But how much is scientific speculation? If all of the global ice melts, the sea may eventually rise by 80 m or more, but if we restrict our view to 2100, the expected rise is up to 1 m. Clearly that has implications for the displacement of populations, and also with respect to international trade as most is carried by shipping that uses ports. Ports are built at sea level with some planned leeway between 'storm + high tide' levels and the top of their quays – but what if these are more often breached; or worse, if ports are totally flooded? International trade will decrease; global gross domestic product will contract; and so on. We should note recent European Union-funded research (Ford, 2013) which focuses on consequences for Europe. This states that sea-level rises are likely to differ around the world, sometimes by tens of centimetres. Damage will be considerable, however. The flood barriers built to protect London, for example, were only expected to be breached once in 1,000 years, but a 1 m rise in the Thames Estuary means that 'you would take that level of protection down from one in 1,000 years down to one in 10 years', said Professor David Vaughan, the programme coordinator. Inevitably, all Asian ports will be affected – those with flat hinterlands will be severely disrupted, while those with mountainous hinterlands will be inundated. Such observations link into the arguments of this book – put simply, 'the situation is complex'. That is what Kidd stressed, and it is what the book's authors bring forward, suggesting that new and subtle decision processes need to be developed – not only for Asian situations but globally.

As Stephen Briggs recently opined (*Financial Times*, 26 May 2013), the simplest systemic failure is that the 'cost of climate' associated with any particular product or service – from the manufacture of a car, to an intensively grown tomato, to an airline ticket – is not inherently included in the price of that commodity. This is the basic disconnect: it is a problem of economics, not science. If the full and true environmental cost of a service were reflected in its price, we could rely on markets at least in part to minimise CC. However, as Martin Wolf (also in the *Financial Times*) said, the effects of CC are perceived to be so far into the future that 'we and us as individuals' will not do anything about it at this time. This argument follows mine above – that governments are so concerned about their daily management that they will postpone discussion and decisions on CC until later, and most of their citizens will concur.

Nevertheless, opinions do change. Let me direct you to the Millennium Development Goals (MDGs) that were agreed in 2000 by all member nations of the UN. The MDGs cover all aspects of humankind development from health and education to CC – and we have globally agreed to meet the goal targets by 2015. These high aims of the MDGs were challenged in 2004 by the Copenhagen Consensus led by Bjorn Lomborg when he brought together a group of economists who, with input from focus groups, created a framework of tasks prioritizing explicitly the world's big problems. The goal was to achieve the most 'good' for people and the planet based on evidence and with limited financial resources. In CC04, as the meeting was called, the four top-ranked projects included three related to health management and one on trade liberalization. The bottom four (of 17 projects) saw little benefit in advancing labour migration or CC programmes, even including the Kyoto Protocol. They were seen as 'not value for money'. Education development did not feature strongly, nor gender issues. However, hunger and malnutrition were top-ranked issues, as well as the promotion of trade. We can appreciate that to make policies stick in any nation an educated workforce is required; and following a better local health policy would yield a cohort of children more able to learn. Lomborg followed up his findings, and in the CC08, results reducing the cost of education rose to issue no. 6, and improving girls' education was no. 8. By CC12, mitigating hunger to aid children's education had risen to the first issue, and deworming children to offer better pathways to health and education rose to fourth. But still CC did not feature as really important. Lomborg has been highly critical of proposals to tackle CC (following Kyoto and its enactments), and he has found that his funding will be cut by the

Danish government: unless new benefactors are found his centre will have to close. While accepting that CC is happening, he argues that current global policies for reducing emissions are not only failing but are a waste of money as well. Instead of trying to cut emissions we should focus on adaptation, he says, investing in renewable technologies and tackling poverty.

It is thus important, as several authors show in detail, to come to (many) understandings of the rationales affecting CC decisions on a case-by-case basis as few are identical one to the other. Their locations differ, as do their history and their forecast future, whatever decision is undertaken. It is thus opportune to see in this book a focus on Asia because it is predicted to be at forefront of the global economic renaissance together with the United States. But to bring the United States into the arguments presented in this book would swamp some of the local issues – size matters. As the old proverb says, 'great oaks from little acorns grow', and thus the examples presented herein illustrate how small regions across Asia can develop results that are crucial for the mitigation of and adaptation to CC.

We will witness changes being enacted which are more recent than those recorded by the authors. Johnathan Kaiman wrote on 30 May 2013 that Shenzhen, China, will experiment with a carbon-trading programme involving 638 companies responsible for 38 per cent of the city pollution as a lead-in experiment to a national total cap on emissions envisaged for the Chinese 13th Five Year Plan (covering 2016–2020). However, this good news masks the fact that China ranks only 116th (of 132 nations measured by the 2012 Environmental Performance Index), but it is trending upwards. And of the other two populous Asian nations, India is doing even worse (than China) at 125th, while the Philippines is performing well at 42nd.

It seems that in making this ecochange, the Chinese government is acquiescing to the pressure of the UNFCCC to become a full signatory to the Kyoto Protocol and its future extensions, but it is a simple fact that in moving rapidly from developing to developed nation it is using the best and next practices to the full – a goal noted by Tek Nath Dhakal and Huong Ha in Chapter 16, though, in the case of the Philippines, they considered that legislation has to be in place. Evidently national rather than international legislation will suffice as the Philippines, and now the Chinese case, emphasise: there may be hope for the globe. Indeed, as I have argued elsewhere (Richter 2002), many Asian businesses have only recently started to focus on issues affecting the broader public, such as CC. A trigger might have been the Asian crisis of 1997–1998 since before then they experienced a period of fast and sometimes unsustainable growth. The Asian crisis was a turning point towards more balanced and long-term-oriented management practices.

All this seems to me to be cause to celebrate the work of the authors of this book in bringing together data and discussions from different regions of

Asia, focused on CC mitigation and adaptation, as well as the requirement for transparent governance. Their work is timely and ought to reach a wide audience.

Frank-Jürgen Richter

Chairman, Horasis, Switzerland

References

S. Briggs (2013) 'So-Called Climate Change Sceptics Are Just in Denial', Letter to the editor, *Financial Times*, 27 May.

Environmental Performance Index (2012) See http://epi.yale.edu/epi2012/rankings. Note several international comparisons at this site.

E. A. K. Ford (2013) *From Ice to High Seas: Sea-Level Rise and European Coastlines*. The Ice2sea Consortium, Cambridge, British Antarctic Survey. See ice2sea@bas.ac.uk.

J. Kaiman (2013) 'China Unveils Details of Pilot Carbon-Trading Programme'. *The Guardian*, 22 May.

J. B. Kidd (2013) 'The "Law of Requisite Variety" May Assist Climate Change Negotiations: A Review of the Kyoto and Durban Meetings'. *Knowledge Management Research & Practice*. Advance online publication 14 January 2013; doi: 10.1057/kmrp.2012.56.

S. Lomborg, see http://www.copenhagenconsensus.com/projects for details of all the Copenhagen Consensus Results.

F-J. Richter (2002) *Redesigning Asian Business: In the Aftermath of Crisis*. Westport Conn.: Quorum Books.

F-J. Richter (2012) 'Let the Big Polluters Lead'. *South China Morning Post*, 11 June.

M. Wolf (2013) 'Why the World Faces Climate Chaos'. Martin Wolf's Exchange, *Financial Times*, 14 May.

Acknowledgements

This book would not have been completed without the support of members of the advisory board, the reviewers, family members, colleagues, the authors and the research assistants. The editors wish to thank Dr Frank-Jürgen Richter (Chairman, Horasis, Switzerland) for providing the Foreword, the reviewers listed below for their professional and constructive feedback, which has been invaluable to the authors, and the authors for their cooperation during the revision stages. Finally, the editors are very grateful for the advice and assistance from the series editor, David Elliott, and, at Palgrave Macmillan, Christina Brian and Amanda McGrath.

List of reviewers

Dr Cristina Mihaela Barbu (Spiru Haret University, Romania)
Prof. Dinesh Chandra Devkota (Tribhuvan University, Nepal)
Prof. R. Lalitha S. Fernando (University of Sri Jayewardenepura, Sri Lanka)
Prof. Gamini Herath (Monash University, Malaysia)
Assoc. Prof. Mohammad Zulfiquar Hossain (Islamic University, Kushtia, Bangladesh)
Major General Md Shafiqul Islam (CRP, Bangladesh)
Assoc. Prof. Ishtiaq Jamil (University of Bergen, Norway)
Prof. Mohammad Mohabbat Khan (University of Dhaka, Bangladesh)
Asst. Prof. Drita Kruja (University of Shkodra 'Luigj Gurakuqi', Albania)
Prof. Sue L. T. McGregor (Mount Saint Vincent University, Nova Scotia, Canada)
Dr Subhakanta Mohapatra (Indira Gandhi National Open University (IGNOU), New Delhi)
Assoc. Prof. Deirdre O'Neill (Monash University, Australia)
Dr S.V.R.K. Prabhakar (Institute for Global Environmental Strategies, Japan)
Prof. A.K.M. Motinur Rahman (University of Dhaka, Bangladesh)
Mr Bimal Raj Regmi (Flinders University, Australia)
Dr Claus-Peter Rückemann (Leibniz Universität Hannover & Westfälische Wilhelms-Universität Münster, Germany)
Asst. Prof. Katsuhiro Sasuga (Tokai University, Japan)
Asst. Prof. Isaias S. Sealza (Xavier University, Philippines)
Dr Vinay Sharma (Indian Institute of Technology, Roorkee, India)
Dr Noore Alam Siddiquee (Flinders University, Australia)
Mr Ashok Kumar Singha (CTRAN, Nepal)
Asst. Prof. A.K.M. Ahsan Ullah (American University in Cairo, Egypt)

Contributors

Editors

Huong Ha is the Academic Coordinator of the MBA and undergraduate business programmes at the University of Newcastle (Singapore). She was the Dean of TMC Business School and Director of Research and Development, TMC Academy (Singapore). She holds a PhD in management from Monash University (Australia) and a master's degree in public policy from Lee Kuan Yew School of Public Policy, National University of Singapore. She was the recipient of a PhD scholarship awarded by Monash University, a Temasek scholarship awarded by the National University of Singapore and a scholarship awarded by the UN University/International Leadership Academy, and many other professional and academic awards. She has many years of working and teaching experience in tertiary educational institutes/universities, manufacturing, marketing research and business consultancy/development in Australia, Singapore and Vietnam. She has been an invited member of the CYBERLAWS 2010, 2011, 2012 and 2013 conference committees, dealing with the technical and legal aspects of the e-society; the International Advisory Board of South Asia Association in Criminology & Victimology (India); the Chinese American Scholars Association Board and many others. She has been a reviewer of many journals and international conferences, such as *Thunderbird International Business Review, International Journal of e-Education, e-Business, e-Management and e-Learning, International Journal of Environment and Sustainable Development*, CYBERLAWS Conferences, *African Journal of Marketing Management, International Journal of Consumer Studies*, Academy of Management Annual Meeting 2010 and IEEE International Conference on Computer Science and Information Technology 2011.

Tek Nath Dhakal is Professor and Head at the Central Department of Public Administration, Tribhuvan University (Nepal). He is also the coordinator of 'Governance Matters: Assessing, Diagnosing and Addressing the Challenges of Governance in Nepal' – a joint research project of the Department of Administration and Organisation Theory, University of Bergen (Norway), and the Central Department of Public Administration, Tribhuvan University (Nepal) (www.pactu.edu.np). He has conducted a number of research projects, including 'Status of the Implementation of Citizen Charter in Municipalities in Nepal', 'People's Trust in Public and Political Institutions in Nepal' and 'NGOs' Role in Livelihood Improvement in Nepal', including Institutional Analysis of Markhu and Hekuli VDC (Nepal). He has published three books and three edited volumes, and two research reports both from Nepal and abroad. In addition, he has contributed around 50 articles published both from Nepal and abroad in local and international journals. He also served as an editor of *Nepalese Journal of Public Policy and Governance*, a journal enlisted by ISSN. He often participates in different national and international forums and conferences, and presents papers. He is also affiliated with a number of academic and professional networks, such as ISPA, ISTR, NAPSIPAG, EROPA, the Finnish Society of Administrative Sciences and the Public Administration Association of Nepal.

Authors

Misa Aoki is currently a PhD student at Kyoto University Graduate School of Agriculture, Division of Natural Resource Economics (Japan). She works on the issues relevant to behavioural changes of small-household farmers due to national or local government policies in Japan and other Asian countries. Specifically, she has identified causes for agriculture policies and sustainable agriculture policies from the viewpoint of microeconomics. Earlier as a researcher at the Centre for Non-Profit Research and Information at Osaka University (Japan), she conducted research on the relationship between social capital and policy implementation. She obtained her master's degree from Kyoto University Graduate School of Global Environmental Studies in 2010.

Dinanath Bhandari is Programme Coordinator for Climate Change and the Disaster Risk Reduction Programme in Practical Action (Nepal). He holds a Master of Science in Forestry degree from Tribhuwan University (Nepal) and combines specialist skills in different aspects of community-based natural resource management, disaster-risk reduction and adaptation to CC. He has been involved in different projects with management responsibilities in Nepal for the last two decades. His work over the last seven years has been in the practice and policy on CC and disaster-risk reduction in mainstreaming community-based approaches to development planning, and he has been involved as an expert in developing national- and district-level policy formation and planning processes. He has written and contributed over a dozen articles, books, guidelines and reports in English and Nepali based on his knowledge and work experience.

Ken Coghill is Associate Professor, Department of Management, and a member of the Academic Board at Monash University (Australia). He is a former Councillor (Rural City of Wodonga, 1972–1977), Member of Parliament (Legislative Assembly, Victoria, 1979–1996), Cabinet Secretary and Speaker. His research and teaching interests include accountability, governance and professional development for parliamentarians. He developed the concept of integrated governance in which the three sectors of society (state, market and civil society) interact and are interdependent, with decisions and actions being subject to fuzzy logic. He teaches governance. He is the author of many journal articles and submissions to public inquiries, and he is a frequent commentator on parliamentary and governance matters. His recent books include *Fiduciary Duty and the Atmospheric Trust* (2012) and *Hear Our Voice: The Democracy Australians Want* (2012).

Somnath Hazra has over 10 years of experience in the forestry sector and was the recipient of an award from IDRC (Canada), looking into forestry and CC. He has in-depth knowledge of forest policy evaluation, MAPs, forest valuation, NTFP assessment and market linkages, and biodiversity. He was a university lecturer in environmental economics and worked in various projects related to forestry, rural infrastructure, social assessment, municipal solid waste and the hazardous waste inventory for the West Bengal Pollution Control Board. He has also worked in JFM implementation and evaluation, the People Biodiversity Register, and traditional knowledge documentation capacity-building of non-governmental organisations, community-based organisations and municipalities. He has gained work experience with international organisations such as OXFAM, Broederlijk Delen (Belgium), MISERIOR (Germany) and the Ford Foundation (United States). He has also worked as a consultant in land acquisition on behalf of the government of West Bengal. He has in-depth knowledge in several areas of specialisation, including advisory services on CC projects, the environment and social impact assessment. He has developed CC mitigation and adaptation plans (forestry sector) for five Indian states (Orissa, Meghalaya, Mizoram, Manipur and Tripura). He has expertise of forestry projects in several different countries. He is a forestry consultant to the government of Kenya.

Gamini Herath is Professor of Economics at the School of Business, Monash University, Sunway Campus (Malaysia). He is also the Director of the Social Economic Transformation Multi-disciplinary Platform. He received his PhD from the University of New England, New South Wales (Australia). Before joining Monash University, he worked at the University of New England as well as La Trobe University and Deakin University (Australia). He has published nearly 70 journal papers in highly recognized international refereed journals, including *American Journal of Agricultural Economics, Australian Journal of Agricultural and Resource Economics, Ecological Economics, Journal of Development Studies* and *Oxford Agrarian Studies*. He has published six books, including *Multi-Criteria Decision Analysis in Natural Resource Management* (2006), *Child Labour in South Asia* (2007) and *Institutional Aspects of Water Management* (2012).

R. Lalitha S. Fernando is Professor of Public Administration in the Department of Public Administration, Faculty of Management Studies and Commerce, University of Sri Jayewardenepura (Sri Lanka). She was awarded a full-time scholarship to pursue her PhD in the Graduate School of Public Administration, National Institute of Development Administration, Bangkok (Thailand). She won the prestigious Commonwealth Academic (internal) Scholarship to pursue a postgraduate diploma in development studies, leading to a master's degree in development administration and management tenable at the University of Manchester (United Kingdom). She gained a BSc degree in public administration at the University of Sri Jayewardenepura (Sri Lanka). She has published a number of papers related to public management and governance in national and international journals. Her research interests include managerial innovation, new public management, public entrepreneurship, public policy implementation and evaluation.

Suvra Majumdar is an electrical engineering graduate, a postgraduate in energy management and a certified energy auditor from the Bureau of Energy Efficiency, Ministry of Power. He has been involved in developing descaled models of CC adaptation and mitigation action framed under the state action plan for CC for the government of Meghalaya, Orissa, Mizoram, Manipur and Tripura, with a focus on low-carbon and climate-resilient infrastructure, livelihood resilience, improved service delivery in health, water and sanitation, food and energy security, clean energy and the environment. He has specific exposure in providing low-carbon technology solutions in the area of urban and rural environment and infrastructure, project structuring and institutional development, and supporting the implementation of adaptation and mitigation action. He has worked in and facilitated successful registration of around 20 clean development mechanism (CDM) projects with the UNFCCC. His other work includes policy development and monitoring and evaluation of the implemented programme for Ministry of New and Renewable Energy and the Ministry of Power, promoting cleaner production and low-carbon technology for the government of Orissa, greenhouse gas and energy audit, planning off-grid electrification and facilitating sustainable development at the community level. He has provided services to a number of internationally and domestically funded projects with agencies like the World Bank, United Nations Development Programme, Department for International Development, Deutsche Gesellschaft für Internationale Zusammenarbeit, International Fund for Agriculture Development, and other funding agencies in various sectors and states.

Reina Mashimo has worked as a project researcher at the Institute for Global Environmental Strategies (Japan) in 2011. During this period she contributed to the review and analysis of historical agricultural policies in Japan. She has studied earth system science, seeking a career in CC, particularly in communication and liaising between the public and private sectors. As a system engineer she has three years' work experience in the information-technology industry in Japan. She volunteered for various national and international science and development events in Japan and has travelled extensively.

Subhakanta Mohapatra is working as Reader in Geography at Indira Gandhi National Open University, New Delhi (India). He was trained as a young scientist on the theme of CC and health. His areas of specialisation are the environment, health and development geography. His present research interest in CC is vulnerability and adaptation to CC with a special focus on marginalised communities.

S.V.R.K. Prabhakar is Senior Policy Researcher at the Institute for Global Environmental Strategies (Japan). He works in CC adaptation and disaster-risk reduction in the Asia-Pacific region. Specific research interests include mainstreaming CC adaptation concerns into development plans and policies, measuring adaptation, identifying and characterising adaptive policies, risk insurance, promoting adaptation in the post-2012 climate regime, and training needs assessment for capacity-building. He has more than 15 years of experience in participatory research and development with international and national research and developmental organisations, such as the International Crop Research Institute for Semi-Arid Tropics, Indian Agriculture Research Institute, CIMMYT-RWC (Rice-Wheat Consortium for the Indo-Gangetic Plains), United Nations Development Programme, National Institute for Disaster Management and Kyoto University, where he was responsible for the conceptualisation and execution of a variety of research and developmental projects covering natural resource management, CC adaptation and disaster-risk reduction. He obtained a PhD in field crop management from the Indian Agricultural Research Institute, New Delhi (India) in 2001.

Bimal Raj Regmi is a doctoral student at Flinders University (Australia). He has more than 14 years of professional experience in natural resource management. His work covers the diversity of development and environment issues in Nepal. He has in-depth knowledge of international and national policies and programmes related to CC mitigation and adaptation. His specific expertise includes understanding the legal and policy provisions for CC adaptation; knowledge of the governance of CC and natural resource management; knowledge of participatory research and development; and methodology and approaches to addressing climate risk and vulnerability. He has contributed numerous publications related to CC adaptation, biodiversity and natural resource management.

Abhik Saha has an M Tech in environmental engineering with more than nine years of experience and comprehensive knowledge of environmental impact assessment, CC adaptation and mitigation in different sectors, corporate environmental management systems, environmental monitoring, and water and wastewater treatment. He is accredited for water pollution, ecology and biodiversity, air pollution, and solid and hazardous waste. He has vast experience in environmental assessment from the perspective of sustainable development. His areas of specialisation are CC-mitigation projects based on CDMs under the Kyoto Protocol, voluntary emission reduction projects, and implementing and auditing management systems as per ISO 14001 and 9001. He is a certified lead auditor for the ISO 14001: 2004 standard. He has also been associated with many advisory projects for CC schemes for sustainable urban management, solid-waste disposal and handling. His expertise includes CC adaptation, implementation of the Kyoto Protocol, CC and capacity-building CDM. He has also worked on various assignments in CC adaptation, biodiversity conservation and assessment projects, and institutional issues in forestry sector. He has worked with different medium and heavy industries in sectors such as asbestos cement, sugar and distilleries, chemical and petrochemicals, pharmaceuticals and bulk drugs, paper and pulp, tanneries and tobacco industries in the waste-management sector.

Ramanie Samaratunge is Associate Professor and Deputy Director, Education, in the Department of Management at Monash University (Australia). She obtained a master's degree from Carleton University (Canada) and a PhD from Monash University. She has also had visiting appointments at the Institute of Development Policy and Management, University of Manchester (United Kingdom), and the Sri Lanka Institute of Development Administration. She has an active research agenda spanning the domains of public management, international management and the management of change, and she has published widely in various local and international journals. She is currently conducting research on 'Integration of skilled immigrants into the Australian workplace' and 'Capacity building and good governance in South Asia'. She has secured a number of competitive grants from AusAID on capacity-building and integrated governance in Sri Lanka.

Daisuke Sano is Knowledge Management Team Leader of ADAPT Asia-Pacific. He has been with the Institute for Global Environmental Strategies (IGES) since 2005 and is currently a director of the IGES Regional Centre based in Bangkok. He has been engaged in various projects, including CC adaptation, biofuels, sustainable agriculture and the impacts of trade liberalisation. He has been working on agricultural/environmental issues for 20 years since he started his career at the Ministry of Agriculture, Forestry and

Fisheries of Japan as a technical officer. He holds a doctoral degree in food and resource economics from the University of Florida (United States).

Isaias S. Sealza teaches evaluation research in the Public Administration Programme of the Graduate School of Xavier University. He has been directing research undertakings, mostly policy-oriented studies, on population and development, poverty, child labour, the environment, natural resource conservation and development, peace and conflict, and health issues, commissioned by national and international agencies, for more than 25 years. He has written, as single author or with collaborators, over 100 research reports, journal articles, books, monographs and conference papers, including 'Community Modernisation, In-Migration, and Ethnic Diversification' (*Philippine Sociological Review*, 1982), 'The Cassava Industry in Bukidnon Province' (*Philippine Studies*, 1986) and 'Mindanao Il Conflitto dei Poveri i movimenti d'ispirazione islamica nelle Filippine: trend attuali e reali prospettive di pace con i cristiani' (*Revista di Intelligence*, 2006). He recently completed a performance evaluation of a dairy-farming programme in Mindanao for Land O'Lakes International Development Foundation and the US Department of Agriculture, and a study of the humanitarian assistance for survivors of the tropical storm Washi for UNICEF. He is currently doing research on failed pregnancies for a regional field office of the Philippine Department of Health.

Mahendra Sethi is an urban-environment specialist and editor for the National Institute of Urban Affairs (NIUA) (India). He has extensive research, academic and professional experience in resource management and development planning. He is pursuing a PhD in cities and CC, specifically focusing on low-carbon development. He previously worked with CES-Jacobs (India). Relevant research studies and projects in the field of CC and urban development include 'Capacity Building, Knowledge Networking and Documentation Support for Indian Cities' with NIUA as a part of Asian Cities Climate Change Resilience Network/ Rockefeller Foundation and State Action Plan for Climate Change for Punjab and Regional Plan for Bhagirathi River Valley, Uttarakhand; and Master Plan for Special Economic Zones, Punjab Development Potential Study, Aden (Yemen). Related publications in the field include 'Alternative Perspectives for Land Planning in Today's Context' (*Journal of Institution of Valuers, RNI No. 17683/69* April 2011) and 'Land as a Resource for Sustainable Urban Development – An Environmental Resource Perspective' (Working papers of 59th ITPI Congress, February 2011).

Md Shafiqul Islam, Major General Ndc, Psc (Retd) is the Executive Director of CRP, Bangladesh. Prior to this appointment he served for 34 years in the Bangladesh Army in various posts. He was the Vice Chancellor of

Bangladesh University of Professionals and also served as the Military Secretary of the Bangladesh Army. During that appointment he also worked as the Chief Coordinator of the Biometric National ID and Voter Registration Project. In that project, 80.11 million vote-eligible citizens' biometric database and multipurpose national identity card creation was completed within 18 months and with the lowest possible costs. For successful completion of the project he was awarded the prestigious International People's ID World Award in 2008 in Milan (Italy). He has a master's degree in defence studies and an MPhil in national security studies from Bangladesh National University, and a postgraduate diploma in management from Bangladesh Open University. He has attended various courses in the United States, United Kingdom, Indonesia and Thailand. He is a graduate of Army Command and Staff College, Camberley (United Kingdom), and Defence Services Command and Staff College, Mirpur (Bangladesh). He is one of the pioneer pilots of Bangladesh Army Aviation. In his earlier career he held a variety of command, staff and instructional appointments. He has also travelled to various countries for training, courses and exercises, and as a member of delegations. He served in the former Yugoslavia as UN Military Observer for one year. He has participated in a number of seminars/workshops and written a number of articles in different journals.

Vinay Sharma is a member of the faculty in the Department of Management Studies, Indian Institute of Technology (IIT), Roorkee, Uttarakhand. He has considerable experience of working with various organizations in the fields of media, information technology and social development. He has around 19 years of experience in the areas of strategic management, business opportunity development, co-creation of value, market prosperity through capability development, marketing, rural marketing, business-opportunity development, brand development, IT-enabled services and teaching. His areas of interest include poverty alleviation and rural market development through business development, market development and technology, and he has designed and proposed a specialised model – 'Affordability for the Poor and Profitability for the Provider' – which has also been published as a book. He contributed an appendix on rural marketing to the 13th edition of Philip Kotler's *Principles of Marketing*. He is an associate and member of the founding group of the Network of Asia Pacific Schools and Institutes of Public Administration and Governance constituted in December 2004, and a member of editorial board of *Journal of Administration and Governance*. He has published and presented around 70 papers, and chaired sessions at national and international platforms. He is a member of the editorial board of reputed journals and is also a member of the academic and advisory councils of several prestigious institutions. He is guiding 12 PhD candidates and has been actively participating in prestigious national- and international-level projects, such as the Ganga River Basin Environment Management Plan, a pan-IIT national project in India, and an international project on building a network on social entrepreneurship, co-creation and innovation with partners from Copenhagen Business School (Denmark), FGV-EAESP (Getulio Vargas Foundation), São Paulo (Brazil), VUZF University, Sofia (Bulgaria), and Institute of Management Technology, Ghaziabad (India).

Amita Singh is Professor at the Centre for the Study of Law and Governance, JNU, Delhi (India). As Secretary General of the Network of Asia Pacific Schools and Institutes on Public Administration and Governance, she has written and co-edited a large number of comparative and interdisciplinary research publications. With her specialisation in public administration and political theory, her publications cover a wide area of studies in administrative reforms, land use, access to justice and gender concerns. Her latest book is *Millennium Development Goals and Community Initiatives in the Asia Pacific* (2013, co-edited with Gonzalez and Thompson). During the 1990s she was part of the 'Status of Women Studies' in association with Hunter College SUNY (United States), and between 2008 and 2011 she conducted an extensive survey on cross-cultural best practice in leadership and governance in association with the Global Innovators Network at the Ash Institute of Democratic Governance (KSG, Harvard). Her association with nine universities in southwestern Australia by virtue of her Australia-India Council Fellowship (2006–2007) has resulted in a volume entitled *A Critical Impulse on e-Governance* (co-written with C. Johnson). The Department of Administrative Reforms and Public Grievance (government of India) initiated the Governance Knowledge Centre under her leadership (2006–2009) at JNU. She has since interrogated the 'best practices methodology' in public administration research and suggested alongside, value calculations for the invisible intangibles embedded in ecosystem analysis. She is on the board of the Ethics Committee constituted by the Ministry of the Environment on 'animal experimentation and alternatives'. She is a member of the National High Level Committee on the Status of Women in India. She has won awards for her public service and contribution to empowerment programmes.

Ashok Kumar Singha has a degree in agriculture from Orissa University of Agriculture and Technology and a postgraduate diploma in business management from Xavier Institute of Management. He is the founding Director of CTRAN. He has written extensively on development planning and governance challenges. He has been involved in the design of Western Orissa and is also a contributing author on the Low Carbon Growth special issue of *India Infrastructure Report* (2010). CTRAN was the knowledge partner to the government of Orissa and Meghalaya for the preparation of a state-level action plan on CC. He specialises in energy and CC policy, governance and institutional development areas. He is the co-author of *The Forgotten Sector* (with Vijay Mahajan and Thomas Fisher) and has contributed chapters to several books. The state action plan was supported by GiZ (Deutsche Gesellschaft für Internationale Zusammenarbeit) as part of the ongoing cooperation with the Ministry of the Environment and Forest.

Izumi Tsurita works at the Ocean Policy Research Institute (OPRF), Japan as a researcher with the Policy Research Group. Before joining OPRF, she worked for the Institute of Global Environmental Strategy, Japan as an associate researcher with the Adaptation Team, Natural Resources Management Group. She graduated with an M Sc from the University of Queensland, Australia in 2009 after obtaining an undergraduate degree from Sophia University, Japan, in 2004 majoring in Global Environmental Law in her first two years and Comparative Culture in last two years. Her major interests are CC, biodiversity and marine environmental issues.

A.K.M. Ahsan Ullah is Associate Director at the Centre for Migration and Refugee Studies and Assistant Professor of Global Affairs and Public Policy at the American University in Cairo. He has two master's degrees and a PhD in migration studies. He has contributed extensively to national and international refereed journals, including *Development in Practice, International Migration, Asian Profile, Asian and Pacific Migration Journal, Journal of Immigrant and Refugee Studies, Development Review* and *Journal of Social Economics*. He has also contributed chapters to a number of edited collections and published twelve books in the fields of migration, refugee, and development studies. In addition to academic work, he has worked for national and international development and research organisation for over eighteen years. He has taught and researched at the Asian Institute of Technology (Thailand); City University of Hong Kong; Centre for Development Research, University of Bonn (Germany); Saint Mary's University; McMaster University; and the University of Ottawa (Canada).

1
Governance Approaches to Mitigation of and Adaptation to Climate Change in Asia: An Introduction

Huong Ha and Tek Nath Dhakal

1. Introduction

Climate change (CC) has become one of the pressing issues in several business and policy debates because it is one of the most serious and prolonged threats to the security and well-being of millions of people across all nations. Shortages of food, energy, water, medicine, healthcare and so forth make up a horrific picture of the adverse effects of CC on our planet. The increasingly powerful and more frequent occurrences of typhoons, droughts, floods, tornadoes, earthquakes and volcanic activities/eruptions have intensified such effects. It should be noted that some of these are man-made but others are natural, so solutions to address the causes are different. While, for many reasons, several nations worldwide have not been seriously pursuing any common agenda to either stop or slow down the process, or mitigate the impact of CC, adaptation to its consequences has absolutely become a real challenge for all relevant stakeholders, including states, the private sector, civil society and individuals in different countries.

Climate change is a major environmental, economic and social issue. The increased amount of anthropogenic emissions of greenhouse gases (GHGs) has made CC a major and costly challenge to climate-vulnerable people and communities. The impacts of CC are more pronounced in landlocked and mountainous countries. South Asia is particularly vulnerable to its impacts. Some of the effects have already been seen, such as droughts, downstream flooding, intense rainfall, shifting of monsoon periods and declining crop productivity. Managing the adverse impacts of CC on people's livelihoods, agriculture, water resources, energy, health and biodiversity is a huge challenge that requires comprehensive national strategies and action plans. The occurrences and reoccurrences of several natural and man-made disasters in the Asia-Pacific region, especially in South and East Asia (and recently in North and South America) suggest an urgent and increased demand for inspection, monitoring, capacity-building, and enhancement of public governance and administration to create greater security for the world's population and the global economy.

Since the impact of CC has gone beyond the conventional governance approaches, and has transcended research fields/disciplines and national boundaries, multifaceted and multi-sector governance approaches are required to effectively and efficiently manage CC and its impact. The absence of good governance, especially in South Asian and Southeast Asian regions, has often been ascribed to many different reasons. Among them are insufficient capacity and lack of commitment of the public sector and other groups of stakeholders, lack of transparency and accountability in policy-making and the implementation processes, and a high level of corruption in the public sector.

Governments in the Asian region, especially in South Asia, are at the juncture of searching for novel ways to improve governance and foster coordinated efforts with other non-state actors, such as the private sector, industry and civil society organisations, including non-governmental organisations, consumer associations, professional associations and community-based organisations. Although technological initiatives, such as green-IT and green-power technologies, have been available and used by many companies in some countries, these initiatives should be adopted by many more industries and countries. In this setting there has been a lengthy and intense debate about how all relevant groups of stakeholders can jointly address challenges associated with CC and environmental degradation, and embrace 'better governance' approaches through knowledge management and exchange, and improved organisational learning. Thus it is essential to investigate and analyse how state and non-state actors can cooperatively contribute to the common cause of improving governance in Asia.

The global mechanisms to deal with CC issues, such as the United Nations Framework Convention on Climate Change and the Intergovernmental Panel on Climate Change (IPCC), under which global negotiations are held, must be enabled to work efficiently and effectively to tackle the growing menace of CC in an equitable and just manner. The Durban Climate Conference and the recent Rio Convention have facilitated some progress in the direction of ensuring the operationalisation of the Green Climate Fund in favour of developing countries, including least-developed countries, and globally binding emissions reductions by 2020. The Kyoto Protocol has been extended beyond its expiry in 2012. Given the urgency of the problem, the ways in which these mechanisms are working need to be streamlined and made more effective to ensure that more concrete results will be achieved. Regional organisations such as the Association of South East Asian Nations and the South Asian Association for Regional Cooperation should work in solidarity and with a sense of purpose to fight the menace of CC in a comprehensive manner. They should exert moral pressure on the developed world and relevant international organisations for deeper and speedier emissions cuts and an increased flow of resources, technology and funds for poor countries to cope with their increasing need to adopting appropriate mitigation and adaptation measures.

The objectives of this book are to (i) revisit the issues and impact of CC in Asia, (ii) examine the preconditions for good governance regarding CC, and the role of state and non-state actors in the governance for CC and (iii) explore different political–legal frameworks of decentralised and participatory state-planning, including the enactment of environmental legislation and impact assessment of various environmental projects.

The 14 chapters are contributed by authors from different countries and from various disciplines, focusing on CC impact and various governance models applied across countries in Asia. They are classified under the three following main themes: (i) issues and impact of CC in Asia (four chapters), (ii) preconditions of good governance and the role of different sectors (five chapters) and (iii) governance approaches for managing CC (five chapters).

2. An overview of the chapters

Part I starts with Chapter 2, 'Climate Change, Vanishing Ecosystems and the Challenge of Achieving Human Prosperity', by Amita Singh. This reflects the severe consequences of CC on current ecological and human systems. Land degradation, water depletion and large-scale deforestation leading to the extinction of several species are only some of the impacts of CC and environmental problems. CC not only induces (i) an increase in the cost of ecosystem services and (ii) a reduction in economic benefits; it also contributes to intensifying poverty. Findings from studies by different researchers, such as Pounds and Puschendorf (2004), Pimm (2007) and Chellaney (2011), and international governmental and non-governmental organisations, such as the World Resources Institute, the United Nations (UN) Environment Programme, the IPCC, the UN and others, confirm that there is a strong correlation between ecosystem depletion and human well-being. Singh highlights that some 'convenient truth' alerted the policy-makers and relevant stakeholders. She goes on to argue that isolated efforts to address issues associated with conservation of some commercial species in order to achieve sustainable development may produce adverse outcomes and jeopardise global progress regarding environmental sustainability. She explains that an anthropogenic or human-centred approach to achieving prosperity raises ethical questions of intragenerational equity and justice. Nevertheless, this situation can be improved considerably or reverted through both incremental and transformational changes in the current governance approaches, including policies, institutions, mechanisms and programmes, which aim to prevent undesirable trade-offs and enhance positive synergies in ecosystem conservation efforts.

Chapter 3, 'The Interplay between Climate Change, Economy and Displacement: Experience from Asia', by A.K.M. Ahsan Ullah, discusses the interdependence between CC, economy and displacement in the context of Asia. He agrees totally with Amita Singh that the impacts of CC on human beings have become one of the most pressing and sophisticated issues that must be addressed not only at the national level but also at the international level. CC can potentially and vigorously increase social and economic costs, cause habitat reduction and trigger human displacement in various locations in Asia. He clarifies that although human displacement has increasingly gained worldwide attention, efforts to address the imminent health and safety concerns of those who are affected by such displacement are not as robust as is merited. This is due to disagreement about how to respond to CC since the parties involved perceive that some responses may pose a threat to their personal or national interests. Therefore it is important for the involved parties to search for a consensus on how the accords and the agreements related to CC can be endorsed and stringently adopted.

Chapter 4, 'Disaster Communication in Mitigating Climate Change in Sri Lanka: Problems and Prospects', by R. Lalitha S. Fernando focuses on disaster communication in mitigating CC in Sri Lanka. This a very important topic, especially during crises. The long-lasting and haunting experience of the 2004 tsunami, and another disastrous situation in 2011, causing considerable damage to people's lives, property and to the economy, are very painful. They are costly lessons of how better communication during disasters can save many lives. Fernando stresses that if citizens are ignorant about possible disasters, their lives are at significant risk, and without communication, no disaster management is possible. Ineffective communication is considered to be one of the main barriers to the management of natural disasters in Sri Lanka. This chapter examines various measures which can help to improve two-way communication in the process of mitigation of CC in Sri Lanka. Fernando argues that information and communications technology can be used by relevant stakeholders as a feasible and effective means to communicate with the public and with each other in the governance process to respond to CC.

In Chapter 5, 'Climate Change and Global Environmental Governance: The Asian Experience', Gamini Herath explores various impacts of CC, not only from the standpoint of an environmentalist but also from that of an economist and a futurist. The key issue discussed in this chapter evolves around the escalated increase in GHG emissions worldwide as discussed by other authors – namely, Beg *et al.* (2002), Shui and Robert (2006), and Raman (2009). He explains that although developing countries have adopted several measures to respond to the widespread and long-lasting adverse effects of CC, insufficient capacity together with inadequate financial and social resources create one of the barriers to effective mitigation and adaptation to CC impacts. This issue does not pertain to developing countries, but developed countries have also struggled in establishing a common understanding of how to govern CC effectively and efficiently. One of the examples cited by Herath is the limited effectiveness of the Kyoto Protocol due to the disagreement among political leaders in respective countries. The leaders of many developed countries have refused to rectify this agreement for many reasons. Ironically, these countries produce the most GHG emissions per capita, but at the same time they call for novel and innovative approaches to prevent and alleviate the destructive effects of CC.

At the start of Part II, Chapter 6, 'Approaches to Climate Change Adaptation: A Case Study of Agricultural Initiatives in Japan', by Izumi Tsurita, S.V.R.K Prabhakar and Daisuke Sano, discusses various measures to adapt CC impacts on the agricultural sector. Unlike others, the authors claim that research studies on how a developed country addresses issues associated with CC is as important as debates and studies on CC adaptation in developing countries. Thus the chapter examines the initiatives taken in the agricultural sector in Japan, using two prefectures – namely, Niigata and Miyazaki, as case studies. The findings reveal that these prefectures have embarked on an ambitious plan to develop rice varieties that can withstand the predicted climatic vagaries. However, such initiatives are limited and fragmented, and they lack central coordination and cooperation. The authors suggest that it is imperative to formulate an all-inclusive masterplan which embraces multistakeholder, multisectorial and multilevel approaches in order to address the ongoing and new challenges of the governance of CC.

In the context of Japan, S.V.R.K. Prabhakar, Misa Aoki and Reina Mashimo explore CC adaptive policies at the macro level in Chapter 7, 'How Adaptive Policies Are in Japan and Can Adaptive Policies Mean Effective Policies? Some Implications for Governing Climate Change Adaptation'. They observe that policies, structures and institutions, which are adaptive in nature or are designed to change, are assumed to be able to cope with CC impacts (Parry *et al.*, 2007). Yet there has been insufficient assessment of the effectiveness of such systems in terms of adaptation to CC due to the unavailability of substantial experience in CC adaptation in most countries. This chapter explores the development of agriculture policies over time in Japan. It also examines the challenges which policy-makers and implementers have faced and addressed during the course of action (Ohara and Soda, 1994). The findings of their study demonstrate that variables, such as the timeliness of introduction of policies and frequent revision of policies, may not be correlated with the success of adaptive policies because other variables – for example, the understanding by relevant stakeholders of the issues on the basis of which the policy was made – also affect the effectiveness of the adaptive policies in the agriculture sector.

In Chapter 8, 'Management of Climate-Induced National Security: Paradigm Shift from Geopolitics to Carbon Politics', Md Shafiqul Islam debates another dimension of CC management – that is, the management of climate-induced national security. He explains that the vulnerabilities induced from CC include human displacement, reduction of clean water, loss of livelihood, health hazards, energy crisis, and change of hydrological patterns and ecosystems. Such vulnerabilities directly affect human security, which eventually leads to national insecurity (Joshua, 2007; Muniruzzaman, 2011). Islam argues that, in order to address the threats to national security associated with CC, there should be a paradigm shift from geopolitics to carbon politics – that is, from the current approaches, such as low-carbon economy, carbon-trading, carbon politics and so on, to new and integrated governance approaches to contain such threats.

Many countries, both developed and developing, have created a national plan to deal with CC impacts, as has India. According to Ashok Kumar Singha, Suvra Majumdar, Abhik Saha and Somnath Hazra in Chapter 9, 'Deconstructing Debate on the National Action Plan on Climate Change at the State Level: A Case Study of Meghalaya State, India', an action plan was introduced in 2008. The authors explain that in order to implement the plan successfully, an action framework is required. They argue that various groups of stakeholders should be engaged in the design and implementation of the framework since national priorities may not address the challenges faced by the stakeholders at the subnational level and may not be consistent with state priorities in terms of responding to climate chance effectively (O'Brien *et al.*, 2004; Thomas and Twyman, 2005). The chapter discusses the process by which several groups of stakeholders at the state level have been engaged in the process of prioritising actions through the adoption of a prioritisation matrix, and have developed a financial plan for the implementation of such actions.

In line with the discussion in Chapter 9, Vinay Sharma also emphasises the importance of stakeholder participation and engagement in environmental governance to respond to CC and other environmental problems in Chapter 10, 'The Role of Government and the Private Sector in Mitigating and Adapting to Climate Change'. According to his experience when participating in a major project which addressed CC impacts in India, the government can formulate CC policy. Yet it cannot take full responsibility for a reduction in economic growth and an imbalance between economic development and sustainability due to the trade-offs between environmental conservation and economic development owing to many constraints. Since the private sector is a major driver of social and economic development, this sector has an important role to play in the governance process to mitigate and adapt to CC impacts (Ha, 2012).

In Chapter 11, 'Integrated Governance and Adaptation to Climate Change', Ken Coghill and Ramanie Samaratunge depict an integrated governance approach which can mobilise all sectors in society. Unlike other studies on governance which focus on structures, mechanisms, institutions, communications channels and so on, this chapter identifies a different set of variables which can significantly influence the governance process, such as belief systems, culture and economic practices. The authors explain that these variables can affect the way in which various sectors in society – namely, the public sector, the private sector and civil society organisations – participate in the integrated governance model to contain the intensification of CC and adapt to its adverse effects (Frey and Stutzer, 2000; Knight and Rosa, 2011; Fox-Decent, 2012).

Another governance approach to mitigating and adapting to CC impacts – a multi-sector governance model – is introduced in Chapter 12, 'Climate Change Governance: The Singapore Case, by Huong Ha. Ha explains that the governance of CC is multidimensional, intricate and multilevel due to the nature of the effects of CC and environmental degradation. Therefore the governance of CC must involve participants from different sectors of society, such as government, business and civil society at the local, state, national and international levels (Al-Amin, Jaffa and Sitar, 2010; Driessen *et al.*, 2012; Falaschetti, 2013). Also, both regulatory and non-regulatory mechanisms should be adopted in the governance process to respond to CC since regulations alone may not apply to all sectors in all contexts and the costs of compliance of regulation may outweigh the benefits gained from it (Fisher and Surminski, 2012). Ha also identifies and analyses the main forces – such as leadership, resources, strategic planning and implementation – affecting the success and failure of the multi-sector governance model, using Singapore as a case study.

It is important to discuss how CC affects the urbanisation process since socioeconomic activities in urban areas have contributed significantly to a large amount of GHG emissions (Satterthwaite, 2009; McCarney, 2012). Mahendra Sethi and Subhakanta Mohapatra, in Chapter 13, 'Governance Framework to Mitigate Climate Change: Challenges in Urbanising India', comment that countries which enjoy a high level of population growth rate, rapid urbanisation and a steady increase in economic development, such as Asian and African countries, have faced several challenges in terms of comprehending and incorporating international frameworks into national frameworks to reduce the amount of carbon emissions. Although there is an understanding among policy-makers, researchers, practitioners and so on that metropolitan areas contribute to multiply CC effects, there is insufficient empirical assessment of whether various models adopted by many countries can effectively mitigate GHG emissions and CC impacts. Therefore the authors recommend comprehensive urban-governance mechanisms to respond to CC in India.

Many researchers argue that it is impossible for any governance approach to be implemented without any funding (Bouwer and Aerts, 2006; Bapna and McGray, 2008). The pressure on funding for CC mitigation and adaptation activities does not spare any countries, either developed or developing, and even less developing countries. Nepal has faced the funding dilemma in environmental governance. Chapter 14, 'Unripe Fruits or Non-Raining Clouds? Climate Change Governance and Funding Dilemma in Nepal', by Mr Bimal Raj Regmi and Mr Dinanath Bhandari, depicts the strengths, weaknesses, opportunities and threats of CC governance in Nepal. The authors explain that although policies to address issues associated with CC are formulated, a lack of formal institutions and an effective financial mechanism have hindered the process of policy implementation. A transformational approach to CC governance, embracing a systematic process and formal financial institutions, is proposed to ensure that responses to CC are harmonised and robust enough to translate policies into action.

Finally, in Chapter 15, 'Environmental Legislation and Action in Polity, Economy and Culture for Climate Change Adaptation: A Case Study of Misamis Oriental Province, the Philippines', Isaias S. Sealza and Huong Ha argue that mitigation and adaptation to CC cannot be performed without environmental legislation in place. In line with the international best and next practice, the Philippines government has passed a number of acts to address environmental issues, and has developed a national Climate Change Response Framework (CCRF) (Lofts and Kenny, 2012). This framework assesses the causal interactions among several social, technological, economic and financial variables which affect the course of action to respond to CC. These variables include social mobility, advocacy by the third sectoral society, technological solutions, financial mechanisms and institutions, behavioural and lifestyle change, and others (Junio, 2008). This chapter aims to assess the national CCRF in the subnational context, using Misamis Oriental province as the subject of examination. The findings suggest that the role of the three sectors in society – namely, the public sector, the private sector and the third sector (civil society) – are interdependent. They have to work closely and cooperatively with each other to effect CC adaptation legislation and policies not only at the national level but also at the local and state levels.

3. Limitations

Limitations are unavoidable in any intellectual work, and the same applies to this volume. The chapters here are limited in number due to space constraints. Also, we cannot include research studies in many countries in Asia. However, the coverage of each chapter is wide enough to include important topics associated with CC impacts and governance issues.

4. Conclusion

Overall, the governance of CC is a dynamic process. Given the multifaceted and multilevel nature of CC management, the governance process should be modified, adjusted and adapted to changing conditions in both the external and internal environments of a country. This process will also be affected by international and regional contingencies. The Asian region has been a nucleus of the world in recent decades, and it will be the core of development and transformation of the world in the next few years. Thus, novel and innovative governance approaches are critical to addressing issues associated with CC and environmental sustainability.

Abbreviations

CC   climate change
CCRF   Climate Change Response Framework
GHG   greenhouse gas
IPCC   Intergovernmental Panel on Climate Change
UN   United Nations

References

A.Q. Al-Amin, A.H. Jaffa and C. Sitar (2010) 'Climate Change Mitigation and Policy Concern for Prioritisation', *International Journal of Climate Change Strategies and Management*, 2, 4, 418–425.

M. Bapna and H. McGray (2008) *Financing Adaptation: Opportunities for Innovation and Experimentation* (Washington DC: The World Resources Institute).

N. Beg, J.C. Morlot, O. Davidson, Y. Afrane-Okesse, L.A. Tyani, A.F. Denton, Y. Sokona, J.P. Thomas, E. Lebre La Rovere, J.K. Parikh and A. A. Rahman (2002) 'Linkages between Climate Change and Sustainable Development', *Climate Policy*, 2, 129–144.

L.M. Bouwer and J.C.J.H Aerts (2006) 'Financing Climate Change Adaptation', *Disasters – Special Issues on Climate Change and Disasters*, 30, 49–63.

P.L. McCarney (2012) 'City Indicators on Climate Change: Implications for Governance', *Environment & Urbanisation Asia*, 3, 1, 1–39.

B. Chellaney (2011) *Water: Asia's New Battleground* (Georgetown: University Press).

P.P.J. Driessen, C. Dieperink, F. van Laerhoven, H.A.C. Runhaar and W. J.V. Vermeulen (2012) 'Towards a Conceptual Framework for the Study of Shifts in Modes of Environmental Governance – Experiences from the Netherlands', *Environmental Policy and Governance*, 22, 143–160.

D. Falaschetti (2013) *Global Environmental Governance: Mechanism Design Lessons from Corporate Governance* (Bozeman MT: The Property and Environment Research Center).

S. Fisher and S. Surminski (2012) *The Roles of Public and Private Actors in the Governance of Adaptation: The Case of Agricultural Insurance in India* (UK: Centre for Climate Change Economics and Policy, Munich Re Programme Technical, and Grantham Research Institute on Climate Change and the Environment).

E. Fox-Decent (2012) 'From Fiduciary States to Joint Trusteeship of the Atmosphere: The Right to a Healthy Environment through a Fiduciary Prism', in K. Coghill, C. Sampford and T. Smith (eds.) *Fiduciary Duty and the Atmospheric Trust* (Farnham, UK: Ashgate).

B.S. Frey and A. Stutzer (2000) 'Happiness, Economy and Institutions', *The Economic Journal*, 110 (October), 918–938.

H. Ha (2012) 'A Multi-sector Governance Model for Environmental Sustainability – Australia Case', in J. R. Barker and R. Walters (eds.) *New Zealand and Australia in Focus: Economics, the Environment and Issues in Health Care* (USA: Nova Science Publishers, Inc.), pp. 35–60.

B.W. Joshua (2007), 'Climate Change and National Security, an Agenda for Action', CSR No 32, Council on Foreign Relations, USA.

R. Junio (2008) 'Balabag Exploration Project: Understanding Stakeholders and Analysing Conflict', paper presented at the Conference on Mining in Mindanao, February 28, Xavier University, Cagayan de Oro City.

K.W. Knight, and E.A. Rosa (2011). 'The Environmental Efficiency of Well-being: A Cross-national Analysis', *Social Science Research*, 40, 3, 931–949.

K. Lofts and A. Kenny (2012) *Mainstreaming Climate Resilience into Government: The Philippines' Climate Change Act* (London: Climate and Development Knowledge Network).

A.N.M. Muniruzzaman (2011), 'Climate Change: Threat to International Peace and Security', *The Daily Star*, August 11, 2011, Dhaka, Bangladesh. Council Secretariat.

K. O'Brien, R. Leichenko, U. Kelkar, H. Venema, G. Aandahl, H. Tompkins, A. Javed, S. Bhadwal, S. Barg, L. Nygaard and J. West (2004) 'Mapping Vulnerability to Multiple Stressors: Climate Change and Globalisation in India', *Global Environmental Change*, 14, 4, 303–313.

K. Ohara and O. Soda (1994) 'The Development of Agriculture and Agricultural Policy and the Change of Views on Farming and Rural Society after World War II', *Sanbon Taihaibutsu Shigen Kino Murasaki*, 12, 167–181.

M.L. Parry, O.F. Canziani, J.P. Palutikof, P.J. van der Linden and C.E. Hanson (eds) (2007) *Climate Change 2007: Impacts, Adaptation and Vulnerability. Contribution of Working Group II to the Fourth Assessment Report of the Intergovernmental Panel on Climate Change* (Cambridge, UK: Cambridge University Press), p. 976.

S.L. Pimm (2007) 'Biodiversity: Climate Change or Habitat Loss – Which Will Kill More Species?', *Current Biology*, 18, 3, 117–118.

J.A. Pounds and R. Puschendorf (2004) 'Ecology: Clouded Futures', *Nature*, 427, 107–109.

H.A. Raman (2009) 'Global Climate Change and Its Effect on Human Habitat and Environment in Malaysia', *Malaysian Journal of Environmental Management*, 10, 17–32.

D. Satterthwaite (2009) 'The Implications of Population Growth and Urbanisation for Climate Change', in J. M. Guzman, G. Martine, G. McGranahan, D. Schensul and C. Tacoli (eds.) *Population Dynamics and Climate Change* (New York: UNFPA; London: IIED), pp. 45–63.

B. Shui, and R. C. Robert (2006) 'The Role of CO_2 Embodiment in US-China Trade', *Energy Policy*, 34, 4063–4069.

D.S.G. Thomas, and C. Twyman (2005) 'Equity and Justice in Climate Change Adaptation amongst Natural-resource-dependent Societies', *Global Environmental Change*, 15, 2, 115–124.

Part I
Issues and Impact of Climate Change in Asia

2
Climate Change, Vanishing Ecosystems and the Challenge of Achieving Human Prosperity

Amita Singh

1. Introduction

1.1 The policy challenge

Poverty is intricately woven with ecosystem and biodiversity losses. The United Nations Development Programme (UNDP) (2012) has effectively emphasised that the achievement of several Millennium Development Goals (MDGs) from Goal 1 of reducing extreme poverty and hunger to the improvement of maternal health (Goal 5), reduction of child mortality (Goal 4) and economic development (Goal 8) face severe challenges due to the deterioration of biodiversity and ecosystem losses. While urban inhabitants may also not be able to escape CC related impacts upon the economy, the 1.2 billion rural people living in abject poverty may suffer irretrievable damage to their lives. Most of the ecosystems which provide food, fuel, shelter, medicines, clean drinking water, grazing for livestock, a variety of crops and disaster mitigation may suffer extinction due to CC. Climate-change-related temperature variability has reduced resource availability due to a loss of capacity of ecosystems to function to their optimum, and an increase in intensity and frequency of droughts, desertification, species depletion, soil degradation and crop failures has reduced livelihood options and the vulnerability of human beings. Studies indicate that the world has already exceeded the desired limit of 2 °C, which was accepted at the Cancun Conference of the Parties to the UN Framework Convention on Climate Change (UNFCCC) (UNDP, 2012).

Since this chapter is based upon an analysis of 'ecosystems' as a unit of environmental degradation and well-being, the term needs epistemological clarification at the outset. 'Biodiversity', 'ecosystems' and 'an ecosystem approach' are used intermittently in this chapter as they are interdependent concepts having the same epistemological origin in the discourse on species conservation. The Convention on Biological Diversity (CBD) attempted to define them during the Rio meeting in 1992.

Biodiversity is defined by the CBD as 'the variability among living organisms from all sources including, inter alia, terrestrial, marine and other aquatic ecosystems and the ecological complexes of which they are part; this includes diversity within species, between species and of ecosystems' (UN, 1992, Article 2). Diversity is thus a structural feature of ecosystems, and the variability among ecosystems is an element of biodiversity. The parties to the convention have endorsed the 'ecosystem approach' as their primary framework for action. The CBD defined an ecosystem as: 'a dynamic complex of plant, animal and micro-organism communities and their nonliving environment interacting as a functional unit' (UN, 1992, Article 2).

CC has emerged as a major cause of the extinction of species. The UNDP's Asia-Pacific Human Development Report 2012 is thoughtfully entitled *One Planet to Share: Sustaining Human Progress in a Changing Climate*. The Asia-Pacific region accounts for nearly a third of all of the threatened species in the world. In the last two years (2008 to 2010), two-thirds of countries in the region have experienced an increase in the number of threatened species – the greatest increase is in India where 99 species have been added to the threatened species list (UN Economic and Social Commission for Asia and the Pacific (UN-ESCAP), 2011). For example, the tropical forests spread around Asia contribute to the livelihoods of more than 1.2 billion people yet the tropical deforestation has been severe, with about 13 million hectares of forest lost per year between 2000 and 2010 due to change of land use, which significantly contributed to the emission of greenhouse gases (GHGs) (UNDP, 2012). Similarly, more than a third of the world's largest cities obtain their drinking water directly from protected forest. Many vulnerable ecosystems, such as cloud forests and coral reefs, will cease to function within a few decades and are likely to effect the global economy hugely (Secretariat of the CBD, 2009). Ironically, the majority of countries of the region are committed to achieving the MDGs by 2015, and also to reduce ecosystem losses by 2020, as a commitment to the CBD, yet the consumption and investment process suggests the opposite due to several negative trade-offs in global decision-making. The CBD states that biodiversity conservation is 'a common concern for all humanity' and an integral part of the development process. This convention is critical to the achievement of the other two Rio conventions, which originated at the 1992 Earth Summit—the UNFCCC and the Convention to Combat Desertification along with the Agenda 21. Notwithstanding the commitments under these three conventions mentioned above, there are five multilateral environmental agreements (MEAs) which have a significant role in ecosystem conservation. The most important is the Ramsar Convention on Wetland Conservation followed by the Convention on Conservation of Migratory Species, the Convention on the World Heritage sites, the Convention on International Trade in Endangered

Species of Wild Fauna and Flora, and the International Treaty on Plant Genetic Resources for Food and Agriculture. In the midst of such a broad framework of ecosystem conservation embedded in a philosophy which entails respect for a non-anthropogenic frame of development, the world's economic forces continue to derail the sustainable path of progress and prosperity for mankind. The need for an ecosystem assessment called for by the UN secretary-general, Kofi Annan, in 2000 in his report to the UN General Assembly entitled 'We the peoples: The role of the United Nations in the twenty-first century' highlighted the relationship between ecosystem degradation and poverty. This effort was scientifically undertaken between 2001 and 2005. In 2005 it was published as a report, which looked at the consequences of ecosystem change for human well-being (Millennium Ecosystem Assessment, 2005).

The pressure on ecosystems has increased manifold in the last few decades. While the world population doubled to 6 billion people and the global economy increased more than six-fold, food production increased by roughly two-and-a-half times, water use doubled, wood harvests for pulp and paper production tripled, installed hydropower capacity doubled, and timber production increased by more than half (Millennium Ecosystem Assessment, 2005). The rise of GHGs and changes in atmospheric temperature have led to the disappearance of ecosystems which replenish the earth's environment and create a sustainable base for human security and prosperity. The combination of habitat loss and CC can block any policy on human welfare and prosperity. Large numbers of species, thus far largely unaffected by human actions, are in danger of extinction from CC (Pimm, 2007). Managing development with the least destruction to the environment has remained a major concern for Asian countries that are rich in biodiversity and whose resources of land and water are substantially shared by non-human species. While a small change in the land-use pattern is likely to displace not just a few humans but complex ecosystems, large-scale urbanisation, commercialisation and consumerism is setting at naught even the prosperity so far achieved by humans. In present times the temperature and weather variability due to CC has increased the vulnerability of these ecosystems, which conserve a large number of species, so there is a need to sensitise policy formulation towards more holistic and inclusive development. CC has raised some fundamental questions about the need for conservation and environmental security in public policy. This chapter suggests that the achieved human prosperity of the last two decades will become counterproductive if it advances in defiance of conservation-driven developmental limitations.

There is enough scientific evidence and justification presented in the Intergovernmental Panel on Climate Change (IPCC) Assessment Reports since the first report of 1990 which states that CC has largely been induced by human influence and global anthropogenic activities. The balance of evidence suggests a discernible human influence on global climate (IPCC-SAR, 1996).

1.2　Research objectives

The objective of this chapter is to critically explore the linkage between global progress and human well-being with the loss of ecosystems, and then to evaluate these losses in relation to CC. The chapter assumes that the present advancement is being driven by the requirements of one single species: humans. Humans are constantly pushing their own prosperity in defiance and in conflict with the requirements and survivability of every other weaker species that exists in nature. The environment has become uninhabitable for a large number of species as their habitats of land, water and forests have been destroyed. While this has reduced ecosystem capacity to provide environmental security to human species as much as to other species, the well-being of people in Asia is seriously threatened. Asian biodiversity has already reached its unsustainable limits due to which the region is already on the verge of a vicious cycle of unsustainable growth and increased contingencies in the flow of goods and services to people. This chapter reaches certain conclusions to establish that any human advancement on a scale of prosperity would be sustainable if it were in consonance with nature and in equilibrium with the lives of other non-human species. This manner and mode of development is all the more significant in times of CC in which the environmental stress upon ecosystems has already increased manifold. One of the main reasons for the need for a more holistic integration of ecosystem conservation in human advancement is that it protects both the biological and the cultural diversity which not only conserves but also empowers all living beings against climate-change-related threats and disasters.

2.　Literature review

Some authors and global research findings have particularly influenced the vision, understanding and arguments which were required to write this chapter. While Madhav Gadgil's *Ecology and Equity: The Use and Abuse of Nature in Contemporary India*, which was co-authored with Ramachandra Guha (1995), serves the direction of conservation politics, it also explores the rise of poverty through environmental degradation. Gadgil's *Ecological Journeys: The Science and Politics of Conservation in India* (2001) is a frontal attack on the processes which commercialise ecosystems and endanger biodiversity. Gadgil's *Traditional Ecological Knowledge, Biodiversity, Resilience and Sustainability*, written with Berkes and Folke in 1995, presents the wisdom of traditional practices which conserve ecosystems and are participatory. Thus the argument that every ecosystem nurtures its own unique community bonding and social life, also strengthens the fact that ecosystem diversity and cultural diversity are parameters of sustainable growth. This work draws upon the indigenous production practices of Asian countries to highlight the role of communities in resource management and ecosystem conservation.

The *Millennium Ecosystem Assessment* produced by the UN Environment Programme (UNEP) and World Resources Institute in 2005 presents an empirical survey of ecosystems and emphasises their relationship to human well-being. In the last two decades of climate-change-related changes the adverse impact upon ground water, land and pristine forest reserves in the Asian countries has undergone massive exploitation due to increased demand for ecosystem services. The report has been able to present a strong argument in favour of these ecosystems as capital assets and so their loss is largely irretrievable and irreversible. Climate change has shown a threatening rise in species extinction and ecosystem loss across the Asian landscape. The worst regions are Indonesia, Pakistan and North Korea. China has successfully increased areas under forestation but most countries continue to be lax towards committed action which could reduce the chlorofluorocarbons in the environment. This laxity has become counterproductive to economic prosperity in the region. This report, when read with the *Statistical Year Book for Asia and the Pacific, Resolving Ecosystem Complexity*, provides clear and comparative guidelines for policy action in terms of both intragenerational inequities and the intergenerational impact of such policies, which have failed to resolve socioeconomic issues of global trends in migration, employment, capacity and skill enhancement which can prevent overuse or misuse of resources by creating an overload of demands in a particular ecosystem reserve.

There are some leading contemporary experts on jurisprudence and the ideas of justice who have successfully taken an ethicolegal angle to justify the rights of non-human species. Martha Nussbaum's *The Frontiers of Justice: Disability, Nationality, Species Membership* (2006) has been very well received by legal thinkers, sociologists and jurisprudence experts. A celebrated leader of jurisprudence studies, Upendra Baxi (2007) has aroused concern about the stereotypical understanding of human rights in many of his writings and the latest, *Animal Rights as Companion Human Rights*, has already sent tremors through present-day infrastructure developers and colonisers of land. Many authors and their writings, such as Henry Bugbee's (1974) *Wilderness in America,* suggest a potent policy framework which would resolve with greater sustainability the issues connected to intra- and intergenerational justice. Baxi has analysed some transhuman/posthuman movements which direct our ethical attention to the protection of the personhood claims for these lifeforms (see Baxi, 2007, Chapter 6, and the literature cited therein). Thus 'personhood' is not so much given as constructed; if so we ought to pose the question as to why it remains so vexed an issue with regard to the rights of non-human species. These authors have epistemologically looked into the idea of justice and have found greater sustainability with the biologically inclusive policies which also take cognisance of the rights of non-human species in the environment. A strong defence emerges in favour of more holistic and participatory resource management in which the rights of all those species which exist in the environment are protected. These recent authors proceed with a critique of anthropogenic human advancement initiated by Rachael Carson but suggest convincingly implementable institutional and policy strategies to reverse the negative trade-offs in conservation politics.

On the need for governance reforms which would incorporate the challenges of CC, Roda Mushkat's study entitled *Globalisation and the International Environmental Legal Response, The Asian Context* (2003) suggests three issues which become important in the Asian context: the inevitability of the smooth integration of individual countries in the global economy; the importance of capacity-building to address financial risks and poverty; and the necessity to realise the crucial role played by civil society. Roda Mushkat has suggested an integrated action framework that links environmental, financial and political action to prevent the vulnerability of people to deprivation, destitution and displacement. Much of the literature cited above takes common ground on the need to address issues of ecological justice through the strengthening of community wisdom, participatory democracy, and global and local partnerships on biodiversity conservation.

2.1   Ecosystems and human prosperity

From the conservation point of view, the isolated protection of species may not provide optimum support to the environment. The 'environment' is a web of relationships and 'it resembles the system's view of nature in modern ecology' (Berkes, Folke and Gadgil, 1995, p. 291). There are a number of scientists, anthropologists and biologists (Forline, 2008; Schmitz, 2007; Clark, 2001) who have written about the need to conserve ecosystems for sustainable human advancement. Berkes *et al.* (1995) have focused on the ecosystem-rich Asian countries and one interesting example highlighted in their work has been found in Indonesia. The traditional systems combined rice and fish culture (*subak*), and wastes from this system often flow downstream into brackish water aquaculture systems (*tambak*), which are polyculture ponds combining fish, vegetables and tree crops. The *subak* itself was often part of a water-temple system and the entire regional rice-terrace irrigation system was often managed as a whole, as in Bali. Thus the combination of *subak-tambak* systems for the combined production of rice, fish and downstream crops is an ecologically sophisticated application. The ecosystem view of the environment treats human and non-human animals, plants and rocks as interdependent, symbiotic and interwoven in a web of relationships. This is overlapping cultural and biological biodiversity prevailing in traditional people also called the 'ecosystem people' by Dasmann (1988) because they have an extensive knowledge of flora and fauna, seasonal changes, life histories, animal and plant behaviour and their habitat preferences. Simply by recombination of their ecosystem understanding about different species they are capable of producing outputs which ensure more sustainable returns to human beings. Human ecology is a product of cultural and biological diversity as ecosystem people evolve in relations to their surrounding environment. The Chinese concept of 'living in harmony with nature' is reflected in *Feng-shui* (land, wind and water) and embedded in the Indian mythological concept of *Panchtatva*, which suggests that the universe and all of its contents are essentially composed of five basic elements comprising *Prithvi* (earth), *Apas* (water), *Tejas* (light), *Maruta* (wind) and *Aakaash* (sky). Both concepts indicate 'co-existence with nature' as against the Western understanding of modernisation as a 'conquest of nature' or the 'human relationship with nature as one of separation and dominance' (Dasmann, 1988, p. 179).

Charles Darwin's *On the Origin of Species* (1859) is assumed to have provided justifications for anthropogenic human progress. The much ignored book *Vestiges of the Natural History of Creation* (1844) by Robert Chambers in St. Andrews, Scotland, mentioned by James A. Secord as one which outsold the *On the Origin of Species*, has denounced theological theories which established racial superiority in human evolution. The idea that new species emerge from the pre-existing ones created a storm, and literature of that era revealed that Charles Darwin's theory of Natural Selection was also much compromised in its publication. The above discussion reveals that a dominant race insists on staying in power through its control of institutional structures, contemporary literature and description of social hierarchies. As a result, anything 'different' meets heated opposition and exclusion from society. The argument presented in this chapter thus strongly contests the notion of 'people first' in development policies as it would not be able to achieve the four essential pillars of development as suggested by Mahbubul Haq (2004) – that is, equality, sustainability, productivity and empowerment. If this were to be replaced by 'ecosystem first' or 'development in equilibrium with nature', human life might progress in harmony with nature, which includes a large number of non-human species which inadvertently and invisibly support and empower each other in a habitat. Hasty and thoughtless planning of developmental policies tends to destroy so much inexpensive wealth that exists in nature even though appropriate conservation planning could have multiplied the returns of progress. This understanding is of greater significance during the period of CC as the extermination of ecosystems due to atmospheric changes and natural disasters is already imminent and one need not add to this.

2.2 Issues associated with climate change

CC has undoubtedly reduced human security and threatened human well-being. The much ignored reality of human progress is that human security largely depends upon environmental security. As countries of South Asia compete with each other to increase their growth rates, climate-change-related threats are underestimated and ignored. Nations consistently overlook integrated and concerted action towards the protection of pristine resources from the offensives of CC. Interestingly, there is no alternative to teamwork and partnerships in the region. Nevertheless, this forms the core of the MDG 8 and sadly also the least achieved one. South Asia is on the verge of an environmental disaster. Williams, Bolitho and Fox (2003, p. 270) found that with every 1 °C increase in temperature, almost 40 per cent of core species would disappear, with a 3.5 °C increase 90 per cent and with a 5 °C increase 97 per cent. Further, many other studies have published their findings during the rising threats of CC in 2004 (Pounds and Puschendorf, 2004; Thomas *et al.*, 2004). These studies have stated that in the worst-case scenario a 7 °C temperature rise would lead to the complete extinction of all species, which would mean starvation, disease and the distressing death of human beings due to the lack of food and medicine, and a rise in nutritional deficiency diseases. Even Mahfuz Ahmed, the principal climate-change specialist of the Asian Development Bank (ADB), highlighted the impending disaster for South Asia in the Doha Meet in December 2012 as the region's 600 million poor will be hardest hit, since their livelihoods still depend on these climate-sensitive sectors (ADB, 2012). The concern at Doha indicated an urgent need for a low-carbon and climate-resilient development or to sustain strong economic growth despite the climate-change-related threats (ADB, 2012). The conservation of ecosystems and retaining indigenous knowledge emerged as a dominant argument in the midst of technological and strategic models adopted to prevent CC.

3. Research framework

This study adopts a multidimensional approach to evaluating intragenerational equity and justice in the context of climate-change-affected ecosystems. These dimensions are significant because CC is not just a product of statistical survey through a cost–benefit analysis but an interdisciplinary study area where the valuation of intangible losses needs to be calculated and requisite policies formulated to prevent degradation or initiate mediation to prevent further loss. Since the gestation period of climate-change-related impacts on the environment is long, nebulous and controversial, the multidimensionality becomes imperative. There is also a fuzzy boundary between losses due to cyclic seasonal changes and those changes which are specifically driven by CC.

This chapter undertakes the system's framework of an ecological understanding of the world of human progress. In doing so it relies heavily upon the integrated framework adopted in some of the basic works of Madhav Gadgil to clarify species-bonding processes and reciprocity pre-existing in nature. This understanding has also resonated effectively in the post-colonialisation governance studies conducted by Gaus (1947) and later Riggs (1961). Gaus (1947) explains:

Such an approach... builds quite literally from the ground up; from the elements of place, soils, climate, location, for example- to the people who live there-their numbers and ages and knowledge and the ways of physical and social technology by which from the place and in relationship with one another, they get their living.

(pp. 8–9)

These studies were conducted when Asia was undergoing a period of decolonisation and development. A committed interdisciplinary group of scholars supported by the UN travelled through China, India, Indonesia, Thailand, the Philippines and a few other West Asian countries. Its findings highlighted the fact that the basic unit of development in these regions demonstrated an interdependence of biological and cultural diversity which justifies the need for a decentralised and bottom-up approach in institution-building. The need for ecology-sensitive governance is still being reiterated to sustain the gains of progress (Quah, 2010).

Fikret Berkes and the *Millennium Ecosystem Assessment* produced by the World Resources Institute in 2005. The secondary data were obtained from the Statistical Year Book for Asia and the Pacific. Some of the core questions raised for the survey report are as follows. How are ecosystems and their services changed due to CC and how have these changes affected human well-being? How can the conservation of ecosystems and their contribution to human well-being be enhanced? 'Ecosystems' are understood as those relatively silent environmental factories which produce oxygen, food, fuel and drugs, notwithstanding a web of relationships which provides cultural, social and scientific services to mankind. It is presumed that the cost of development would be much lower and more lasting if we were ready to conserve ecosystems across the world. The conceptual framework of this chapter suggests that humans are an integral part of ecosystems and that a dynamic interaction exists between them and other constituents and inhabitants of ecosystems. These ecosystems provide important recharge-shed for water and for human well-being. They carry the intrinsic value of species and ecosystem. To analyse the criteria of justice in conserving ecosystems, the idea embedded in Hume's *Circumstances of Justice* extends the discourse about justice beyond strict moral law to 'whatever is morally beneficial and humanly possible' (Hume, 1740, p. 488). Kant (1781) would take the conservation of ecosystems as a model of pure case of the rule of law. Following the natural law tradition in jurisprudence, in his *Critique of Pure Reason*, Kant (1781, p. 397) argues that the natural physical world can never be understood exclusively in terms of physical causality governed by exceptional laws because living beings are far beyond that. Kant (1781) explained further that abstract laws cannot be used to explain human–ecosystem relationships. Humans despite being the 'stronger' species should respect the disadvantaged third-party rights or the rights of every non-human species that cannot represent themselves in courts for justice (Chandrachud, 2009).

4. A holistic paradigm of governance for ecosystem conservation

4.1 The need for the conservation of species

An indubitable scientific fact is that human progress largely depends upon the conservation of every other species of plant and animal. Earth is composed of integrated ecosystems which constitute the web of life. Policy-makers have been conserving water, minerals, coastlines and trees, but in a disjointed manner despite the need for convergence and synchronisation of all developmental activities taking place over mountains, rivers, oceans and forest areas (Berkes *et al.*, 1995). Thus segregating items from the whole web of life creates an illusion of development which is short-lived. The targeted attainment of MDGs is being worked out by respective governments in South Asia within the parameter of segregated and secto-rial planning. As a result of this, many poverty-reduction programmes have been implemented in confrontation with nature (land and water) and tend to demonstrate unsustainability. The findings of an epoch-making report prepared by the World Resources Institute in 2005 highlighted the close links between ecosystem degradation and human well-being which cannot be ignored in economic calculations. As ecosystems are destroyed through land and forest conversion, water depletion, and atmospheric and marine pollution, there is a rise in certain diseases, crop resistance, healthcare costs, insurance costs, administrative expenditure, migration, energy wastage and consumer extravagance. The report has substantiated through worldwide surveys that the rise in gross national product (GNP) should subtract the losses in ecosystem services provided by the environment to establish the real picture of human well-being.

Ecosystem losses can never be reversed. In the 1960s after the massive failure of the laboratory-prepared tomato crop in the United States as a result of disease, scientists had to go back to the Amazonian forests to look for the wild berry bushes from where they had extracted the original gene to produce the disease-resistant tomato (Tatum, 1971). The M.S. Swaminathan Trust (1988) found that the promotion of the 'basmati' species of rice over the land of tribals in Orissa (India) made them lose at least 20 species of wild rice which grew in arid and dry weather without dependence upon dam-based irrigation and fertilisers. This not only landed them in debt and malnutrition but finally caused them to be thrown off their land by rich farmers who were able to invest. There are many such recorded findings which have reiterated the demand for ecosystem-based development which is more sustainable and inclusive.

The findings reached in this chapter are not new; scholars have been insisting since the UN-declared development decade in the 1960s and 1970s that the existence of smaller, non-human-driven ecosystems on earth are an indicator of its health and prosperity. The idea was projected in the writings of several philosophers and scientists led by powerful authors such as Aldo Leopold (1949), Henry Bugbee (1962, 1974) and Rachael Carson (1962). The great US advancement led by machines, chemicals and deforestation was challenged for the first time in the report of the Club of Rome entitled *Limits to Growth* (1972). Many leading experts in public policy literature (Nussbaum, 2009) have come forward to defend the rights of non-human species on earth as a significant aspect of planetary justice and inclusive progress. Normative concern surrounds the inability to deliver justice to those that have no speech which could be understood by the mainstream law-makers. As this creates hierarchies of differential abilities in speech, mobility, expressions, language and voice, it also helps us to understand that these differentials exist as much within human species as between human and non-human species. Thus the idea of justice cannot be moulded in accordance with a specie's location in the hierarchy and destroy the very epistemic beauty of being fair and reasonable. Henry G. Bugbee in *Wilderness in America* (1974, pp. 614–620) remarks: 'Wilderness, it would seem, may lie closer to the whence of speaking than to the thematisation of a speaking about...', and he also puts forth an important idea which can form the basis for further discussion of ecological justice.

If wilderness may yet speak to us and place us as respondents in the ambience of respect for the wild – for nature as primordial – it must be liberated from ultimate subsumption to human enterprise – that is, its voice will be heard anew only as we come in decisive forbearance into its presence. Attentive listening, active receptivity and candour of spirit make up the mood of the place (Bugbee, 1974).

In policy debates, 'ecosystem' and 'ecology' are two sides of the same coin as both indicate a constant interaction between the living organism and its environment. While an ecosystem is about communities interacting within themselves, the dimensions of ecology are wider as it in turn is the study of these ecosystems. Once the polarised idea of approaching CC policies is abandoned, one needs to understand closely how ecosystem conservation becomes indispensable to overcome knowledge gaps. The promotion of the philosophy of American exceptionalism (Lipset, 1997) ignored the enormous wealth possessed by the Asian countries in the form of non-cash reserves – that is, their ancient cultural resources of science, literature and medicine, but above all a community bonding which evolved over millions of years in a habitat called a village. Instead of revitalising these pre-existing resources as these nations emerged out of colonial subjugation, the politics of aid implanted new models over them. Comparative studies were less practised during that time, and when F.W. Riggs (1961) preferred to take up this non-West challenge, he did not receive much encouragement from the UN.

4.2 Ecology in governance discourse

Ecosystems as a collection of human and non-human species evolve over a period of time. This generates a form of mutual interdependence based upon a set of unwritten laws of interaction among species. There emerges a form of bonding governed by laws which are loosely referred to as 'ecological' as they emerge to fulfil the need for sharing a common space in a sustainable and peaceful way. John M. Gaus (1947) was one of the first scholars to use the term 'ecological approach' in an otherwise considered 'nuts and bolts' discipline of public policy.

An ecological approach to public administration builds then quite literally from the ground up; from the elements of a place-soils, climate, location, for example, to the people who live there-their numbers and ages and knowledge, and the ways of physical and social technology by which from the place and in relationships with one another, they get their living.

(Gaus, 1947, p. 81)

The idea behind the approaches of Gaus and Riggs was to view non-Western systems in their organic totality rather than in the context of standard and structured development indicators formulated in the West. Since then the last 50 years have raised new demands over the globe and the governance of environmental resources has produced anxieties about the inadequacies of applying the US experience to Asia. The fact that theoretical research has vacillated around the US experience (Singh, 2005) has failed to generate holistic and symbiotic policies of development. Wen-shien Peng (2008, p. 214) writes that 'a polarised model is inadequate in depicting the characteristics that contribute to a developing country's administrative system'. As a result, Riggs (1964) abandoned models that differentiated between agrarianism and industrialism. The 'fused-prismatic-diffracted' model of Riggs (1962) has evolved in contrast with the linear models developed in the Western administrative sciences. It would be difficult to understand the cultural diversity of Asian countries without its reference to its ecology and ecosystems, but post-colonial development and the 'catching up with the West' policies of development have played havoc with the region's land, water, forests and rivers.

Public administration research conducted as part of the comparative administration group generated the need for a holistic paradigm of administration. These concerns expressed in the 1950s, especially those raised in Professor Walter Sharp's September 1952 subcommittee on comparative administration at Princeton University, have insisted on an integrated framework (Sharp, 1953) which made amply clear that public administration suffered deeply from its confinement within Western boundaries or, as Alfred Diamont suggests (1964), 'western experience' (p. 83). Even though this well-intentioned effort was shelved for various reasons, the idea had already sparked wider research in the discipline and by 1959 was reflected as a theoretical model in Siffin's *Toward the Comparative Study of Public Administration* (Siffin 1957), which could transcend Western boundaries and study the systems of developing countries. The case-study method used by Siffin surveyed five developing countries out of which three were from the Asian region – that is, Egypt, the Philippines and Thailand. The ancient irrigation systems in Iran and Sri Lanka are good examples of the engineering ingenuity of early Asia. Despite a good attempt it nevertheless indicated the need for larger samples to overcome the deficits of theory-building. It is ironic that even though this idea infused one of the best-supported movements in public administration and inspired some of the greatest efforts in theory-building, it could not be sustained for more than half a decade. With the coming of CC discourse, this idea of a holistic, ecological or organic approach to public administration returns to haunt the policy-makers.

Humans are merely one species in the vast ecology of life on the planet. However, the share of resources that we extract has created an imbalance in the environment, which is further exacerbated by CC. While efficient states are those that deliver, they are also states which grow in contravention to ecosystem conservation. Thus as a state's capacity to deliver increases, it also evokes greater fear that destruction will be faster and distribution more authoritative. The economic principle known as the Weingast paradox suggests that 'the fundamental political dilemma of an economic system is this: 'A government strong enough to protect property rights and enforce contracts is also strong enough to confiscate the wealth of its citizens' (Weingast, 1995, pp. 1–16). This indicates a serious governance crisis embedded in the developmental paradigm of global economic systems which need to be addressed if CC is to be reversed.

5. Habitat and ecosystem services

5.1 Assessment habitat and ecosystem losses

Shrinking green habitats, loss of water bodies and land degradation have severely challenged the sustainability of human prosperity. In saying so this chapter may reiterate or repeat the anguish of Rachel Carson's *Silent Springs* written in 1962 but in the present decade it is not simply about the extinction of bees and butterflies to precede an epidemic of scarcity but about an extinction of human life from the surface of earth and that too in a most bizarre manner, as indicated by the tsunami in Asia and hurricanes in the United States. Water is the primary ecosystem resource Asia is the world's most challenged water scarcity area (Chellaney, 2011). It has the world's lowest per capita access to freshwater. The unbridled extraction of groundwater and lack of any committed policy for improving recharge-sheds of water has been turning the region into a drought-prone area. However, the high growth rate and substantial rise in per capita income mean that food production alone consumes more than 92 per cent of the water reserves. This region also has the largest number of dams, numbering around 50,000, and a little more than half of these are in China alone. Salween, for example, is Asia's last free-flowing river, originating in China and flowing through the great Himalayan range through Tibet, Myanmar and Thailand till it finally concludes its journey in the Andaman Sea. The dams destroy its diverse biological ecosystems (Chellaney, 2011).

The transboundary basins of South Asia have been benefiting from the four major river basins – namely, Ganges–Brahmaputra–Meghna, China–India–Nepal, Indus–Pakistan–India–Afghanistan and the Helmand Basin at the intersections of Afganistan–Iran–Pakistan. There is no scarcity of water if it is appropriately managed and the annual water withdrawal statistics in the region also suggest that much is possible with regard to water conservation. The highest annual water withdrawal of 84 per cent is for agriculture, 10 per cent for industrial purposes and only 6 per cent for domestic purposes. The loss of water bodies and the subsequent drying and desertification of land has resulted in a severe loss of resources that support human and environmental security. Some 57–70 per cent of mangroves, which protect the shores and coastal areas, have been lost in the last 30 years of development, and this has brought more than 1.6 million tons of sediment into the Indian Ocean annually (UN-ESCAP, 2011). Fisheries as a means of livelihood are being wound up by coastal communities due to a huge loss of fish, shrimp and aquaculture for food, which is calculated at more than 60 per cent in this region.

It has been shown that the loss of wetlands across Asia has largely been due to their conversion into rice-growing fields. Some 90 per cent of the 130 million ha of paddy land is in Asia (Millennium Ecosystem Assessment, 2005). Some of the most severely damaged regions in Asia are the Red River delta in Vietnam (1.75 million ha), the Sylhet Basin in Bangladesh (1 million ha), Central Myanmar (6 million ha), the Central Indian Plains (40 million ha) and the Central Plains of Thailand (1.9 million ha). The world had already destroyed most of its wetlands in the 1990s and this triggered a CC impact on the atmosphere. Scott (1993) offered a worrying picture of Asia's loss of wetlands, which was 31 per cent in Indonesia, but the loss of mangroves in Singapore, the Philippines and Thailand was as high as 97 per cent, 78 per cent and 22 per cent, respectively. The species-extinction rate has also increased. The *Millennium Ecosystem Assessment* (2005) suggested that the extinction rate was up to 1,000 times as high as the previous extinction rates.

An imbalance in resource use has also contributed to the loss of ecosystems. Almost 90 per cent of water withdrawal in the region is for agricultural use (UN-ESCAP, 2011), and this diverts developmental investments into megaprojects that destroy the local ecology and result in the silting and salination of rich agricultural land. The IPCC Fourth Assessment Report indicates the changed pattern of rainfall. There is a greater intensity of rainfall while the duration becomes shorter. This would leave little water for croplands and result in greater losses through flooding. The IPCC anticipates reduced availability of freshwater in large river basins which would affect more than 1 billion people in Asia by 2050. On the expected rise in temperature to 3 °C, the retreat of glaciers and thawing of permafrost may bring severe damage to ecosystems but more than that would be the lengthening of the heatwaves in Asian countries (Cruz et al., 2007).

The *Statistical Year Book for Asia and the Pacific* (2011) has given the rate of destruction of forest areas as being the highest at 30 per cent in Pakistan and North Korea, whereas Indonesia has almost destroyed 241,000 sq. km during the present decade. The loss of mangroves and coastal ecosystems has increased human vulnerability to CC upheavals in the atmosphere and oceans, energy-related emissions constitute nearly two-thirds of the global total, and 60 per cent of emissions are generated in developing Asia Pacific (UNEP, 2011). In South Asia, carbon intensity increased gradually during the 1980s and early 1990s, but since the late 1990s it has been declining, with the gross domestic product growing faster than emissions (World Bank, 2011).

The 2010 figures of the Forest Resources Assessment of the Food and Agriculture Organisation (FAO) show that 34 per cent of the world's forestland is primary, compared with 25 per cent in Asia and the Pacific (FAO, 2010; FAO and JRC, 2012). The largest regional stocks are situated in the Russian Federation, Indonesia, Papua New Guinea, India, China and Thailand. The average annual growth rates of forest areas in the South Asian countries during 2000–2010 show that there has been a sharp decline in Nepal, Bangladesh, Sri Lanka and Pakistan, while good results have come from the Maldives, Bhutan and India (FAO and JRC, 2012). Carbon emissions are greatly increased due to deforestation besides the magnitude of emissions from degradation is estimated at 30 per cent of total emissions from the forest sector globally (TNC, 2009).

Protected marine area are a rich source of biodiversity but economically backward nations have not been able to invest in appropriate conservation plans. Australia and Kiribati are leading the Asia-Pacific countries with 28 and 20 per cent of their marine areas protected, respectively. Even this falls short of the Strategic Plan for Biodiversity (2011–2020) objectives adopted in 2010 in the 10th meeting at Nagoya, Japan, the primary objective of which is to 'improve the status of biodiversity by safeguarding ecosystems, species and genetic diversity' (UN-ESCAP, 2011, p. 97).

5.2   Ecosystem services and economics

Ecosystem services are the conditions and processes through which natural ecosystems, and the species that make them up, sustain and fulfil human life. They maintain biodiversity and the production of ecosystem goods, such as seafood, forage timber, biomass fuels, natural fibre, and many pharmaceuticals, industrial products and their precursors (Daily, 1997).

Applying an economic tool of analysis to assess the value of goods and services provided by a production unit, one can make a reasonable assessment of the value of ecosystems. The *Millennium Ecosystem Assessment* (2003) refers to three types of ecosystem service that directly benefit peoples' well-being:

- 'providing products (for example, food, fuelwood, medicine);
- regulating services (for example, water purification, climate or erosion control); and
- cultural services (for example, recreation, spiritual, religious)' (MEA, 2003, p. 57).

During the last two decades the increased pace of habitat modification and invasion of pristine ecological regions, and the increased emissions of GHGs due to land-use changes, have affected both species and ecosystems.

The population of the Asia-Pacific region is expected to grow by 600 million between 2010 and 2020, reaching a total of 4.2 billion people (FAO, 2009). This is going to increase the rate of land clearance for agricultural and infrastructural expansion, followed by the setting up of hydroelectric and thermal power stations to fulfil the energy needs of people, production and prosperity. Between 1990 and 2010 the forests of Southeast Asia contracted by 3.32 million ha – an area greater than that of Vietnam (FAO, 2010, p. 1). The 2011 UN-ESCAP (2011) report highlights the fact that Asian and Pacific countries accounted for almost half of the world's carbon dioxide emissions in 2008, which is a very steep increase from their share of 38 per cent in 1990. The climate-linked shifts in temperature and precipitation determine the survivability of ecosystems and the varieties of local livelihoods that are dependent upon them.

Lately, many infrastructural and commercially viable strategies of introducing invasive species are proving to be seriously disastrous for natural habitats and groundwater. The IPCC (2007) has mentioned such activities as anthropogenic drivers of growth as they create inequities and environmental imbalance simply to promote the prosperity of a few powerful human beings.

A recent study conducted at the National Center for Ecological Analysis and Synthesis published in the 2 May 2012 issue of *Nature* found a close correlation between CC, biodiversity losses and reduced plant growth. Ecologist Bradley Cardinale of the University of Michigan who cowrote the paper warns that loss of biodiversity due to species extinctions will bring major changes on earth. The study involved a database drawn from 192 peer-reviewed publications which looked at ecosystems and species variabilities. The findings suggest that an ecosystem loss of 21–40 per cent of species is expected to reduce plant growth by 5–10 per cent (Hooper *et al.*, 2012).

The *Millennium Ecosystem Assessment* (2005) sounds the death knell of two ecosystem services – capture fisheries and freshwater, which are now well beyond levels that can be sustained. Climate change has triggered a sea-level rise beyond natural and adaptable cycles of nature. It is estimated to be almost 5 mm per year for the next 100 years, which requires serious policy changes in the coastal regions of Asian countries, especially to protect ecosystem zones of mangroves and natural forests.

The UN Millennium Declaration of 2000 emphasised sustainable water-management strategies at the regional, national and local levels that would promote both equitable access and adequate supplies. Water has a much larger connection to the lives of the poor. The Human Poverty Index developed by the UNDP in 1997 suggests access to safe water as a key indicator of poverty, which complements the Human Development Indicators of well-being (UNDP, 2006).

The *Asia Pacific Yearbook*, however, laments that the road to reach Nagoya objectives remains long (UN-ESCAP, 2011). CC has hastened ecosystem and species extinction as indicated by the claim of the IPCC (2007a, 2007b). According to IPCC estimates, more than 20–30 per cent of species globally fall under a high risk of extinction by 2100 as global mean temperatures rise by 2–3 per cent.

6. Ecosystem management and a framework of justice

6.1 Human well-being and ecosystem conservation

Human well-being and CC are inversely dependent because the latter destroys the ecosystems which nourish the former. Ecosystem preservation is the bridge between governance and environment. An economically insufficient government can also take measures which improve well-being through strategic planning for ecosystem services and monitoring their sustainable delivery.

As mentioned in the previous section, the lives of poor people are intricately interwoven with their environment from where they draw resources for their livelihood, such as food, fuel, house-building materials, indigenous crops, drugs and occupational resources. Environmental changes which degrade any of these subsistence resources end up increasing their vulnerability, notwithstanding the rise in GNP for the mainstream population. Dasmann (1988) has named them 'the ecosystem people'.

Life is more vulnerable and insecure to face the wrath of nature in the form of increased frequency of disasters and human losses. Between 2001 and 2010, natural disasters killed an average of more than 70,000 people every year in Asia and the Pacific and more than 200 million people were affected in some way every year. Of the world total the Asia Pacific region included 90 per cent of those affected by natural disasters, 65 per cent of deaths due to natural disaster and 38 per cent of economic damage from natural disasters during that decade (UN-ESCAP, 2011). The annual report of the UNEP (2012) highlights the losses due to disasters in the region for 2011. It says that 2011 was a year of natural disasters in which an estimated loss of more than US$366 billion was incurred. More than 29,700 lives were lost in a total of 302 disasters, but Thailand suffered almost two-thirds of the damage to lives and property.

Incorporating CC mitigation measures in governance has also been helpful in sustaining well-being. The differentiated data suggest that while the high-income countries in Asia-Pacific suffered more than the developed countries of the North, yet the relative impact was less in comparison to that on the low-income countries of the same region. While in the high-income Asia-Pacific countries only about 1 in every 1,000 people were affected by disasters and 1 in 1 million died annually, in low-income countries nearly 30 in 1,000 people were affected and 52 in 1 million people killed (UN-ESCAP, 2011). Sadly, alternative approaches are still not being adopted in all countries in Asia despite the realisation that the environment sees no boundaries.

Organisation for Economic Co-operation and Development (OECD)/ International Union for Conservation of Nature (IUCN) 1996 guidelines for conservation have indicated that about 46 million people per year are at risk of flooding due to storm surge and land losses. Predictions regarding losses due to sea-level rise are almost 17 per cent for Bangladesh alone, whereas 90 per cent of the beaches in Japan will disappear. Changes in the frequency and intensity of precipitation will affect the magnitude of runoff, flood and drought. El Niño and La Niña will continue to occur, resulting in recurrent droughts and floods. Studies on the Philippines by Raddatz (2009) also present a threatening scenario. The temperature and inundation of the Lake Lanao watershed in Mindanao, Angat Dam in Northern Luzon and coastal areas in Manila Bay will rise. This will affect more than 2,000 people and 5,000 ha of land in Manila, notwithstanding severe storms in the highly populated areas there.

6.2 Mitigation measures and the search for justice

Environmental justice debates could easily be handled in the pre-climate-change era when the simple 'polluter pays principle' could be enforced upon erring parties. It was simpler to make an assessment about intra- and inter-generational equity and justice when forests were not looked upon as storehouses or carbon dioxide sinks, or rivers had fewer riparian obligations in regions with political dissensions and antagonisms. As environmental justice graduated to 'value calculations' rather 'price alone', as in tort law, the understanding of embedded subjectivity in calculating the intangible resources and the methodology of calculating the cost and damages became more and more complicated. Nevertheless, unravelling this complexity through the study of species interdependence and calculating ecosystem services through the tools of ecological economics becomes a reality of understanding sustainable human well-being.

The problem of justice is one of finding sufficient information about the victim and the perpetrator in the struggle for depleting resources during CC. One would be better equipped to deal with justice issues in the light of indicators of consumption patterns, speed of land-use changes, deforestation, depletion of groundwater, increasing city slums, drinking water availability and laws related to environment management. The principle of scarcity evokes an understanding of how equitable distribution could be in the developmental process.

The warning given by Edward A. Page (2007, p. 175) is meaningful. He argues that the 'Kyoto Lite' approach is being driven by the US administration under the rubric of its Clear Skies, Global Climate Change and FutureGen initiatives, and seeks to put voluntary targets for national emissions based on the ratio of national carbon emissions to economic output. He (2007) suggests that the negative trade-offs being made to obtain voluntary targets should be offset by sufficient incentives to businesses in technology transfer of developing countries.

The UN has taken a number of initiatives to reverse CC impact upon ecosystem degradation. In collaboration with the World Resource Institute and Oxfam, the Caring for Climate Initiative and Adapting for a Green Economy: Companies, Committees and Climate Change have been introduced. These initiatives have become one of the largest voluntary business and CC initiatives. Bangladesh and Bhutan in South Asia have reduced the vulnerability of their populations by taking reasonable measures in environmental governance (UNEP, 2011). In 2011 another scheme called the UNDP-UNEP-Poverty Environment Initiative was set up to help nations to manage their ecosystems more effectively.

The *Millennium Ecosystem Assessment* (2005) suggests a number of issues in dealing with the issue of just ecosystem management. Some insights arrived at in the report are discussed below. Principally, an important way to restore justice in ecosystem management is to prevent negative trade-offs. If land-use changes are indispensable then a proper study of the interdependence of local ecosystems could be carried out. Perhaps this would test our traditional understanding of how to increase the flow of one service without reducing the flow of others, which are likely to be affected in the process of making changes. Many pro-poor resources available in the common property areas or dependent upon these areas, or the waste land for graziers and peatland for many other village communities, are income-generating.

A change of land-use policies ought to be made more sensitive towards these grassroots issues which are close to the less privileged inhabitants of land. These may also culminate in many positive trade-offs which could enhance the sturdiness of ecosystem services to people. Second, an application of ecosystem-specific technology is also a challenge in the provision of justice. Collective choices may become more political rather than rational as a result of which ecosystem services may not be equitably distributed among various groups in society. Public choice theorists have warned against the nature of the modern liberal state, which enhances access of the powerful groups to vulnerable ecosystems despite the warning of unsustainability and embedded environmental disasters. This could be contained through the use of appropriate technology for a particular ecosystem and by restricting its control to the hands of an institution rather than an interest group.

The design of institutions becomes the lighthouse of ecosystem management. One need not suggest a monolithic or monoculture of institutional design. This has been continuously attacked by pro-people groups across the world. These groups are constituted of aboriginals, tribals, forest dwellers, anti-dam activists, village folks, slum dwellers, farmers, seed protectors, cattle herders, tree sitters, wilderness groups, archaeological conservation societies and wildlife protectors. Decision-making processes, decisional priorities, jurisdictional issues, biological requirements and cultural diversity issues vary across countries and regions. Keeping this in mind, environmental governance may recommend a culture-specific institution where the indigenous wisdom and experience of the seniors could be absorbed into the institutional mechanism.

Third is the design of tools for decision-making. This suggests that an appropriate legal framework to guide the decision-makers and methods will help it resonate with the most justified requirement of sustainability. By using tools for increased participation through capacity-building, education and exposure to information about the nature of environmental resources can empower ecosystem communities to manage their ecosystem resources and mobilise forces against unsound policy-making.

To summarise, a three-pronged approach is required for ecosystem management. This matches the suggested action by UNEP (2011) in its regional e-newsletter for the Asia-Pacific: (i) capacity-building, training and sensitisation towards indigenous solutions for ecosystem management through which a holistic planning process could be adopted; (ii) participatory, accountable and transparent institution-building for ecosystem management and (iii) appropriate use of technology, priority-setting and budgeting techniques to prevent the degradation of resources.

7. Limitations

Climate can be assessed through remote-sensing, geographical information systems and meteorological establishments in every country in the world.

Yet, none can admit that the information provided would match the current level of preparedness, thus the damage can be reduced but never reversed. This is as much an issue of governance as it is of the accuracy of deciphering information. Undoubtedly countries need better technology and improved models for the assessment of information.

Ecosystems are organic entities and an understanding of their nature is limited. A scientific study of the conditions, trends and evolution to sturdier lifespans of ecosystems may end up providing incomplete information. Also there are not enough inventories, databanks and projections on the behaviour of ecosystems to ensure the accuracy of assessment and conservation techniques.

There exists a complete lack of partnerships between local resource groups and international institutions which could resonate with each other. Global institutions are still not prepared to admit grassroots wisdom which is not commercially viable. This creates a disjunction between the globally and locally assessed value of ecosystems and their services. This has also prevented an understanding of the interdependence of biological and cultural services. Thus many ecosystems are lost because an understanding of them is located with communities which are governed somewhat differently.

There is also a fuzzy separation between the degradation of an ecosystem and CC as its cause. The causality is embedded into a debate on the existence of CC itself. This is exacerbated by the interest groups spread around the decision-making processes to control the process and also the share of resources from ecosystems. In such a game the state may disclose, divert or conceal information about the value of pristine biodiversity areas. Coalition-based, politically unstable and undemocratic military states are most likely to hide information so as to serve a dominant interest group.

8. Conclusion

CC is a living reality and there is no escape from it. Asia has the world's richest biodiversity regions and also half of the world's population. It also has the largest number of poor who are 'ecosystem dependent' for food, fuel, water and many other services. The accent on luxury and consumerism in Asia turns the tables against their development and progress as their bank of resources which is the ecosystem is drained out for the benefit of the cities' rich populations. This chapter does not suggest an anti-growth and anti-comfort paradigm which it may be mistaken for but recommends an improved governance approach for ecosystem management. CC is most likely to affect the poor who are the low-energy-consuming, ordinary people. Their lives are intertwined with ecosystems and a large number of non-human species that cannot be evaluated on an economic scale of cost–benefit analysis. This chapter suggests an 'ecosystem approach' to the management of biodiversity. This is possible through environmental governance where institutions become knowledge-driven, are able to generate capacity for the participatory appraisal of their decisions, transparent in information dissemination and driven by the demands of inter- and intragenerational justice.

Abbreviations

ADB	Asian Development Bank
CBD	Convention on Biological Diversity
CC	climate change
ESCAP	Economic and Social Commission for Asia Pacific
FAO	Food and Agricultural Organisation
GHG	greenhouse gas
GNP	gross national product
IPCC	Intergovernmental Panel on Climate Change
IUCN	International Union for Conservation of Nature
MDGs	Millennium Development Goals
MEAs	multilateral environmental agreements
OECD	Organisation for Economic Co-operation and Development
UNDP	United Nations Development Programme
UNEP	United Nations Environment Programme
UN-ESCAP	UN Economic and Social Commission for Asia and the Pacific
UNFCCC	United Nations Framework Convention on Climate Change

References

ADB (2012) 'Climate Change Threat to South Asia', http://www.adb.org/features/climate-change-threat-south-asia, date accessed 3 December 2013.

U. Baxi (2007) *Human Rights in a Posthuman World: Critical Essays* (Delhi: Oxford University Press).

F. Berkes, C. Folke and M. Gadgil (1995) 'Traditional Ecological Knowledge, Biodiversity, Resilience and Sustainability', in C. A. Perrings, K. G. Mäler, C. Folke, C. S. Holling and B. O. Jansson (eds.) *Biodiversity Conservation* (Netherland:Kluwer Academic Publishers), pp. 281–299.

H. Bugbee (1962) *The Tale of a Wood* (Canada: Knopf).

H. Bugbee (1974) 'Wilderness in America,' *Journal of the American Academy of Religion*, 42, 4, 614–620.

R. Carson (1962) *Silent Springs* (London: Hamish Hamilton).

A. Chandrachud (2009) 'From Culture to Cow Urine: The Moral Paradoxes of India's Hindu Right', *Harvard Law Record*, March 19.

B. Chellaney (2011) *Water: Asia's New Battleground*, (Georgetown: University Press).

R. Chambers (1844) *Vestiges of the Natural History of Creation*, (London: John Churchill).

T. W. Clark (2001) 'A Course on Species and Ecosystem Conservation: An Interdisciplinary Approach', in T. M. Clark, M. Stevenson, K. Rutherford and

K. Ziegelmeyer (eds.), Bullettin No. 105, *Species and Ecosystem Conservation: An Inter-disciplinary Approach* (New Haven, Conn: Yale School of Forestry and Environmental Studies).

R. V. Cruz, H. L. Harasawa, S. Wu, Y. Anoxhin, B. Punsalmaa, Y. Honda, M. Jafari, C. Li and H. N. Nguyen (2007) 'Asia Climate Change 2007: Impacts, Adaptation and Vulnerability', in M. L. Parry, O. F. Canziani, J. P. Palutikof, P. J. van der Linden and C. E. Hanson (eds.) *Contribution of Working Group II to the Fourth Assessment Report of the Intergovernmental Panel on Climate Change* (Cambridge, UK: Cambridge University Press), pp. 469–506.

G. C. Daily (1997) 'Introduction: What Are Ecosystem Services?' in G. C. Daily (ed.) *Nature's Services: Societal Dependence on Natural Ecosystems* (Washington, D.C: Island Press), pp. 1–10.

R. F. Dasmann (1988) 'Towards a Biosphere Consciousness' in D. Worster (ed.) *The Ends of the Earth: Perspective on Modern Environmental History* (Cambridge, UK: Cambridge University Press), pp. 177–188.

C. Darwin (1859) *The Origin of Species*, (London: John Murray).

FAO and JRC (2012) *Global Forest Land-Use Change 1990–2005, FAO Forestry Paper No. 169* (Rome: FAO of the UN and European Commission Joint Research Centre).

FAO (2009) *State of the World's Forests 2009* (Rome, Italy: FAO), ftp.fao.org/docrep/fao/011/i0350e/i0350e01b.pdf, date accessed 28 Feb 2013.

FAO (2010) *FAO Forestry Paper 163* (Rome: FAO of the United Nations and European Commission Joint Research Centre), www.fao.org/docrep013/i1757e/i1757e.pdf, date accessed 28 Feb 2013.

L. C. Forline (2008) 'Putting History Back into Historical Ecology: Some Perspectives on the Recent Human Ecology of the Amazon Basin', *Journal of Ecological Anthropology*, 12.

J. M. Gaus (1947) *Reflections on Public Administration* (Alabama: University of Alabama Press).

M. Gadgil, R. Guha (eds.) (1995) Ecology and Equity: *The Use and Abuse of Nature in Contemporary India*, (London: Routledge).

M. Gadgil (2001) *Ecological Journeys: The Science and Politics of Conservation in India* (New Delhi: Permanent Black).

M. Haq (2004) 'The Human Development Paradigm in Sakiko Fukuda-Parr', in A. K. Shiva Kumar (ed.), *Readings in Human Development, Concepts, Measurement and Policies for a Development Paradigm* (New Delhi: Oxford University Press).

D. U. Hooper, E. C. Adair, B. J. Cardinale, J. E. K. Byrnes, B. A. Hungate, K. L. Matulich, A. Gonzalez, J. E. Duffy, L Gamfeldt and M. I. O'Connor (2012) 'A Global Synthesis Reveals Biodiversity Loss as a Major Driver of Ecosystem Change', *Nature*, 11118, 2 May 2012 (online) *Nature* 486, 105–108.

D. Hume (1740) *A Treatise of Human Nature* (London: C. Corbet).

IPCC (2007a) *Climate Change by Susan Snell Solomon, The Physical Science Basis: Contribution of Working Group I to the Fourth Assessment Report of the Intergovernmental Panel on Climate Change* (Cambridge, UK: Cambridge University Press).

IPCC (2007b) *Climate Change 2007: The Physical Science Basis. Contribution of Working Group I to the Fourth Assessment Report of the Intergovernmental Panel on Climate Change, S. Solomon, D. Qin and M. Manning* (Cambridge UK: Cambridge University Press).

IPCC-SAR (1996) *Climate Change 1995: The Science of Climate Change.* Contribution of Working Group 1 to the Second Assessment Report of the IPCC (Cambridge UK: Cambridge University Press).

E. Kant (1781) *Critique of Pure Reason* by J. M. D. Meiklejohn (London: Henry G. Bohn), 1829 translation.

A. Leopold (1949) *A Sand County Almanac, with essays on conservation fromround river.* (USA: The Random House, Ballantine Book).

S. M. Lipset (1997) *The American Exceptionalism, A Double Edged Sword* (USA: Norton).

Millennium Ecosystem Assessment (2003) *Ecosystems and Human Well-being: A Framework for Assessment* (Washington D.C.: Island Press).

Millennium Ecosystem Assessment (2005) *Ecosystems and Human Well-being: Synthesis* (Washington D.C.: Island Press).

R. Mushkat (2003) 'Globalisation and the International Environmental Legal Response, The Asian Context', *Asia Pacific Law and Policy Journal*, 4, 1, 1, 49–81.

M. C. Nussbaum (2006). *The Frontiers of Justice: Disability, Nationality and Species Membership* (USA: Harvard University Press).

M. C. Nussbaum (2009) *The Therapy of Desire, Theory and Practice in Hellinistic Ethics* (New Jersey: Princeton University Press).

OECD/IUCN (1996) *Guidelines for the Aid Agencies for Improved Conservation and Sustainable Use of Tropical and Sub-tropical Wetlands* (Paris: OECD).

E. A. Page (2007) *Climate Change, Justice and Future Generations* (Cheltenham UK: Edward Elgar).

W-S. Peng (2008) 'A Critique of Fred W. Riggs' Ecology of Public Administration', *International Public Management Network*, 9, 1, Electronic journal at http://www.ipmr.net.

S. L. Pimm (2007) 'Biodiversity: Climate Change or Habitat Loss – Which Will Kill More Species?', *Current Biology*, 18, 3, 117–118.

J. A. Pounds and R. Puschendorf (2004) 'Ecology: Clouded Futures', *Nature*, 427, 107–109.

J. S. T. Quah (2010) *Public Administration-Singapore Style* (UK: Emerald Group Publishing).

C. Raddatz (2009) *The Wrath of God: Macroeconomic Costs of Natural Disasters, Policy Research Working Paper No. 5039* (Washington, D.C.: The World Bank).

F. W. Riggs (1961) *The Ecology of Public Administration* (Bombay: Asia Publishing House).

F. W. Riggs (1962) 'Trends in the Comparative Study of Public Administration, *International Review of Administrative Sciences*, 28:1, 9–15.

F. W. Riggs (1964) *Administration in Developing Countries, The Theory of Prismatic Society* (USA: Haughton Mifflin Company).

O. J. Schmitz (2007) *Ecology and Ecosystem Conservation* (Washington D.C.: Island Press).

D. A. Scott (1993) 'Wetland Inventories and the Assessment of Wetland Loss: A Global Overview', in M. Moser, C. Prentice and van Vessem (eds.) *Waterfowl and Wetland Conservation: A Global Perspective in the 1990s* (Slimbridge, England: International Waterfowl Research Bureau Proceedings of a Symposium in Florida, USA).

Secretariat of the Convention on Biological Diversity (2009) *Connecting Biodiversity and Climate Change Mitigation and Adaptation: Report of the Second Ad Hoc Technical Expert Group on Biodiversity and Climate Change*, Technical Series No. 41 (Montreal: Secretariat of the Convention on Biological Diversity).

W. Sharp (1953) *Committee on Public Administration, Subcommittee on Comparative Public Administration, 'Final Report' Sept.* Reprinted in The Sayre-Kaufman Outline, A Research Design for a Pilot Study in Comparative Administration (USA: CAG Occasional Papers), p. 2.

W. J. Siffin (1957) 'Toward the Comparative Study of Public Administration', in: W. J. Siffin (ed.), *Toward the Comparative Study of Public Administration* (Bloomington: Indiana University Press).

A. Singh (2005) 'Indian Administrative Theory: Context and Epistemology', *Administrative Theory and Practice*, 27, 1.

M. S. Swaminathan Trust (1988) 'Sustainable Food and Nutritional Security for the 1990s', *Agricultural Situation in India*, 42, 5, 363–367.

L. A. Tatum (1971) 'The Southern Coral Reef Blight Epidemic', *Science*, Volume 171. *An Interdisciplinary Approach* (New Haven: Yale University).

TNC (The Nature Conservancy). (2009) Noel Kempff Mercado Climate Action Project: A Case Study in Reducing Emissions from Deforestation and Degradation. Washington: TNC.

C. D. Thomas, A. Cameron *et al.* (2004) 'Extinction Risk from Climate Change', *Nature*, 427, 145–148.

UNDP (2006) *Human Development Report* (New York: UN), http://hdr.undp .org/hdr2006/pdfs/report/Human_development_indicators.pdf, date accessed 5 June 2013.

UNEP Regional Office for Asia Pacific (2011) *Ecosystem Management*, www.unep.org/ road/activities/ecosystem management/6057/2013, date accessed 5 June 2013.

UN-ESCAP (2011) *Statistical Yearbook for Asia and the Pacific* (Thailand: ESCAP Statistics Division), http://www.unescap.org/stat/ data/syb2011, date accessed 5 June 2013.

UN (1992) *Convention on Biological Diversity*, File name; Ch_XXVII_8, Vol.2, Rio de Janeiro, http://www.cbd.int/doc/legal/cbd-en.pdf, date accessed 5 June 2013.

UN Development Programme (UNDP) (2012) *One Planet to Share: Sustaining Human Progress in a Changing Climate.* Asia Pacific Human Development Report (London, New York, Delhi: Routledge, Taylor and Francis).

B. R. Weingast (1995) 'The Economic Role of Political Institutions', *Journal of Law, Economics and Organization*, Vol. 11. No. 1 Spring.

S. E. Williams, E. E. Bolitho and S. Fox (2003) 'Climate Change in Australian Tropical Rainforests: An Impending Environmental Catastrophe', *Proceedings of the Royal Society of London*, 270, 1527, 1887–1892, 22 September 2003.

World Bank (2011) The Changing Wealth of Nations: Measuring Sustainable Development in the New Millennium (Washington DC: The World Bank).

3
The Interplay between Climate Change, Economy and Displacement: Experience from Asia

A.K.M. Ahsan Ullah

1. Introduction

Only after the devastating tsunami of 2004 shook the policy-makers, scientists and academics has climate change (CC) emerged as one of the strongest forces that can cause economic loss and massive human displacement. CC could now be linked to a number of contemporary dimensions, such as securitisation, poverty, migration and refuge. However, climate security or environmental security poses a few questions. Whose security, what do we mean by security and what are those elements that pose threats? Have we ever thought that climate and climatic disruption may pose threats to human kind that seriously? The attendant consequences of climatic disorder, such as crop failure and infrastructural breakdown, may lead to severe poverty conditions which may tend to social unrest (Garnaut, 2008). Both poverty and unrest can create sufficient conditions to induce population displacement. Until recently, migration and refugee scholars might have somehow missed seeing the link between CC and displacement. The Indian Ocean disaster in 2004, the Fukushima catastrophe in 2011 and some other environmental tragedies beyond Asia, such as Haiti and Chile, suggest that there is a direct link between CC with human displacement. The classical migration theorists tended to generalise myriad factors of displacement using a push–pull approach.

The devastating climate consequences that have taken place in the last two decades and the possible threat of the disappearance of most of the small island developing nations, such as Kiribati, Vanuatu, Tuvalu and Samoa, in the coming four to five decades have shaken the conscience of world leaders. Kiribati will disappear in 50 years, and Tuvalu is negotiating the purchase of land for its citizens from Australia and Fiji. The Maldives is already lying below sea level. A cabinet meeting of Mohamed Nasheed's government was held under water in order to attract the attention of the world to the imminent danger that the country is expected to encounter. The relevant question to ask here is what the economic value of a nation state is. CC has been a significant agenda at the global level for a long time; however, the issue was rarely viewed from an academic angle (Daron, Johnson and Robinson, 2001; Reuveny, 2007; Tol, 2009). Moreover, CC and its attendant consequences, and the interplay between human displacement and economy, have come into this domain as a policy matter recently. However, serious attention has not been paid to such emerging issues by global policy-makers, academics, think tanks or civil society organisations.

Is CC a natural response, an anthropogenic one or both? Scientists have presented overwhelming evidence that CC, in a negative way, has been contributed by human activities (Holzmann and Jorgensen, 2000; Intergovernmental Panel on Climate Change (IPCC), 2007a; Ullah, 2012a, 2012b). The facts about emissions of greenhouse gas (GHG) are important in this discourse. This is, however, a more technical, political and scientific discourse than a social science one (Crompton, Pielke and McAneney, 2011). Since the onset of the discourse about CC and migration, there have been attempts at both global and regional levels to manage CC and its resultant effects. This chapter aims to analyse how effective those attempts are. Enforcement of these attempts depends largely on the goodwill, commitment and sincerity of the respective governments. This chapter goes on to argue that the direct association between CC, economy and displacement has come to the fore in the global development agenda in general and in Asia in particular.

2. Literature review

CC is a global concern, so measures should be adopted globally and regional and domestic actions should be complementary. Which countries, regions and continents are vulnerable to CC? The answer is the 'entire world'. However, regionally or continentally, the severity and forms of CC vary. For example, the African continent is more vulnerable to drought than floods while the Asian continent is the opposite (Heltberg, 2007; Heltberg, Siegel and Jorgensen, 2009). The implications of CC encompass the social, political and economic landscapes both regionally and internationally.

Today, global CC is one of the most dangerous challenges that the international community faces. Since CC is likely to have profound effects on agriculture, settlement patterns, natural disasters, disease and economic activity, future scenarios and potential human impacts can easily be imagined. There is no doubt that CC would exacerbate resource scarcity, which would lead to social unrest and conflict, and these, in turn, would create human dislocations (Cullen and Sarah, 2007; Idean, 2008), and these always come with huge economic losses.

CC may have both positive and negative impacts depending on the geographical location involved. For example, in countries with extreme (low) temperatures, such as Canada, Russia and some parts of Europe, an increase in temperature by 2 °C or 3 °C would be a benefit as a result of higher agricultural yields, lower winter morbidity and lower heating costs (Driessen and Glasbergen, 2002). Obviously, extreme cold weather is not favourable for tourism, so this change would help the countries to boost tourism. Developed countries in lower latitudes are not spared either by the fact that water availability and crop yields (e.g. in Southern Europe) are expected to decline by 20 per cent with a 2 °C increase in global temperature (Baer and Athanasiou, 2007; Asian Development Bank (ADB), 2009a, 2009b; Fung, Lopez and New, 2011).

Nonetheless, all of the countries that are vulnerable to these negative consequences are in developing regions. Africa and Asia, especially Sub-Saharan Africa and South Asia, are at the biggest risk from CC (Heltberg, Siegel and Jorgensen, 2010). Meanwhile, rich nations, as mentioned above, will not be spared either. For example, a 1 m rise in sea level may affect 13 million people in five European countries and destroy property worth $600 billion (Adger, 2006; IPCC, 2007c; Ananthaswamy, 2009). The top 10 countries at risk from the impact of CC, in order of their CC Vulnerability Index (CCVI), are Haiti, Bangladesh, Zimbabwe, Sierra Leone, Madagascar, Cambodia, Mozambique, the Democratic Republic of the Congo, Malawi and the Philippines (IPCC, 2007a; Anthoff, Hepburn and Tol, 2009). One-third of them are from Asia. The CCVI singled out six cities in the world that are at extreme risk from CC impacts. They are perceived as the fastest-growing cities in the world, such as Calcutta in India, Manila in the Philippines, Jakarta in Indonesia, Dhaka and Chittagong in Bangladesh, and Addis Ababa in Ethiopia (United Nations (UN), 2012).

CC is directly affecting biodiversity, ecosystems and resource bases, and both directly and indirectly affecting humans. Drought, cyclones, wildfire and storm surges are the greatest risks posed by CC (see Figure 3.1). Droughts and wildfire in Australia, Russia, the United States and Africa, and floods in Bangladesh, Pakistan, Sri Lanka and Central America, are obvious consequences of CC (IPCC, 2012). Various studies assume that 1–3 billion people will experience water scarcity in the coming seven decades and 200–600 million will suffer hunger. Also, 2–7 million will experience coastal flooding every year (Christian Aid, 2007). The Christian Aid agency further predicts that in the coming four decades about 1 billion people will be displaced from their habitat by global warming. The repeated accentuation exerted by Myers (2002, 2005) on the estimated 50 million people who had migrated by 2010 and the number of people displaced by CC in China was 30 million (Myers, 2002, 2005).

The world will witness three times as much drought by 2070 and seven times as many floods and cyclones by 2070. In fact, it is striking to see

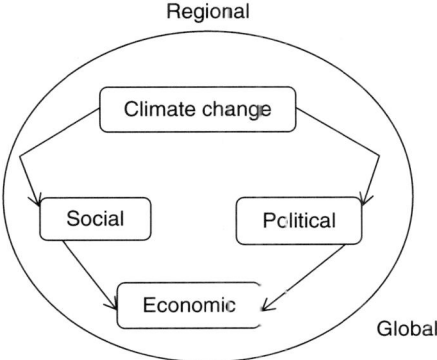

Figure 3.1   Impact of climate change at regional and global levels
Source: By the author (2012).

that the changes in GHG in the last 100 years have been tremendous. For example, CO_2 (carbon dioxide) increased by 34 per cent, CH_4 (methane) by 152 per cent, N_2O (nitrous oxide) by 18 per cent and chlorofluorocarbons by 0–880 parts per trillion (US Department of Transportation, 2010). Worldwide, 'desertification is making approximately 12 million hectares of land useless for cultivation every year'; about 29 per cent of the globe's surface is land (Food and Agriculture Organisation of the UN (FAO), 2009, p. 2). Out of this, the FAO (2009) estimated that 39 million km^2 (26 per cent) were 'forested land' in 2000. Also, 15 million hectares of forest are lost every year (FAO, 2009).

Changes in the polar ice sheets could raise sea levels by 1 m or more by 2100 (Poore, Williams and Christopher, 2000) and the implications could be severe. The more ice and glaciers melt, the higher the rise in sea level. The implication is that about 10 per cent of the world's population lives in vulnerable coastal areas, and 75 per cent of vulnerable people are in Asia. More striking is that, globally, 1.2 billion people (23 per cent of the world's population) already live within 100 km of the coast and in the coming two decades this number is expected to increase to 50 per cent (UN Educational, Scientific and Cultural Organisation (UNESCO), 2009).

Not only are Asian countries at the risk of sea level rises but many European countries are at the same level as the sea, and the Netherlands is already below sea level. If global average warming exceeded 1.9 °C for long enough, the Greenland ice sheet would melt, resulting in a sea-level rise of about 7 m. This is a scary warning because a 1 m rise in sea level means that 60 million people would be living within 1 m of the mean sea level. It is expected that this number will rise to 130 million by 2100 (IPCC, 2007b; Nicholls *et al.*, 2007).

The primary concern about CC and migration is that the latter continues to result from complex combinations of the classical push-and-pull model, underlying causal dynamics and triggering events. CC drives along other factors rather than fundamentally shifting existing contexts and characteristics of migration (Ewing, 2012). There has always been a sweeping tendency to associate migration instantly with economic factors which has overshadowed many other underlying factors. For instance, intraregional migration within any region could not be explained by economic models alone. What are the reasons for Sudanese people to move to Eritrea or for Eritreans to move to Burkina Faso and so on? In Asia, apparently, intracountry mobility takes place due to climatic reasons, generally from coastal areas inland. However, exceptions are not rare indigenous population and Afar population in Ethiopia are perfect examples.

Most variables behind catastrophic consequences, such as, drought, desertification, sea-level rise, deforestations and so on, began to surface long ago. All of these consequences are reminiscent of the predictions that scientists made earlier. More alarming is that during the coming half-century, drought-affected areas will continue to expand. In Asia, China appeared as a significant country in the debate about CC because it did not sign the Kyoto Protocol and it emits arguably the largest amount of GHG, putting a huge population in a highly vulnerable position.

Southeast Asia is often perceived in the climate literature as a hotspot of CC-induced migration, in part because it is already 'migration active', with increasing internal mobility and cross-border migration (Ducanes and Abella, 2009; Ullah and Hossain, 2011). One-third of the Southeast Asian population is subject to CC-induced risk, and these are located in low-lying coastal areas of Indonesia, Myanmar, the Philippines, Thailand and Vietnam. This fact is important and needs to be further analysed in order to compare the situation with South Asia. The latter suffers a high level of poverty and poor governance (Duit, 2008), which has a major impact on its adaptation strategies for CC. This implies that CC and poverty combined induce migration from South Asia (Ullah, 2010). CC compounds the existing poverty in Asia. Their dependence on natural resources and their limited capacity to adapt to a changing climate put Asians into vulnerability to poverty. Those who have the fewest resources and the least capacity to adapt to CC are the most vulnerable. Severe events in CC threaten livelihoods hugely. Although agriculture contributes about 2.5 per cent of global gross domestic product (GDP), its share is much larger in Asian countries (Hertel, Burke and Lobell, 2010). The farm sector in the broader economic performance of poor countries in Asia is significant, therefore the potential macro- and microeconomic effects of CC on agricultural production has crucial implications. Studies confirm that low agricultural productivity in countries of Asia and Africa is the direct contribution of CC (Hertel *et al.*, 2010).

3. Conceptual framework

It is now an established fact that climatic disorder is linked directly to population migration. The phenomenon has freshly occupied space in the discourse of how to manage climatic disorder to mitigate human suffering. It was recently acknowledged that CC contributes to the displacement of habitation and economic loss. It is predicted that more than 200 million people will be displaced by the effects of CC over the next four decades (Hugo, 2008; IPCC), 2012; Ullah, 2012b).

This chapter, however, takes the straightforward position that CC has an impact on human displacement and this displacement does not come without economic loss. Economic loss may cause and be caused by displacement. Obviously, it is not an easy task to measure this loss in economic terms nor is there any method currently available to do so. We may find anecdotal estimates of human loss but no such assessment was made of the economic loss incurred from the Tsunami disaster, the Fukushima catastrophe or the Haiti tragedy. However, at this level it is possible to explore the direct and indirect link between CC, economic loss and human displacement.

In fact, there are so many gaps in our knowledge of CC, migration and the relationship between them in Asia and the Pacific that it would be irresponsible to precisely offer an estimate of the number of people likely to migrate in the future. This study does not intend to repeat the work by the IPCC and others. Rather, this chapter draws upon the key issues relating to the impacts of CC. These could be viewed from three levels – namely, (i) the primary level, which results directly from changes to climatic patterns, (ii) the secondary level, which refers to changes to environmental systems resulting from primary impacts and (iii) the tertiary level, which refers to the broader impacts on societal systems, including implications for migration patterns, and implications of potential policy responses by governments. There are a number of ways in which CC may affect countries, such as exposure, impact sensitivity, adaptive potential and capacity. Conceptually, exposure refers to the likelihood of vulnerability to CC effects, based on current metrics. Impact sensitivity demonstrates the magnitude of disruption resulting from these impacts. Adaptive potential refers to the economic resources available to a country to manage its vulnerability. These key factors determine which country is at most risk and its ability to cope with the climatic disorder (Alberini, Chiabai and Muehlenbachs, 2006) (see Table 3.1).

The relationship between CC and population migration is a widely recognised reality, though relatively little empirical research has been done so far to delve further into the details of specific regions. Speculation suggests that it remains one of the hardest tasks to determine the likely origins or destinations and the likely number of people who may be motivated to move due to CC.

Table 3.1 Vulnerability indicators

Indicators	Summary rationale
Exposure	
Average temperature	A higher starting average temperature indicates greater vulnerability
Temperature changes	A higher rate of increase in average temperature suggests greater vulnerability to changing weather factors
Water availability	A lower water availability per capita value indicates greater vulnerability to CC factors
Water availability	A higher negative percentage change of renewable water per capita indicates greater vulnerability
Extreme events	A higher level of extreme events indicates greater exposure
Change in extreme events	A higher rate of events indicates an increasing magnitude of climate risk
Impact sensitivity	
People affected	More people affected reflects greater vulnerability
Deaths	More people affected reflects greater vulnerability
Damage costs	Higher damage costs as a proportion of the economy reflects greater vulnerability to CC-driven weather events
Adaptive potential	
Wealth	A lower GDP per capita indicates greater vulnerability because of the lower ability to invest to adapt
Budget	Higher debt indicates a lower capacity to pay for infrastructure build
Adaptive capacity	
Rule of law	Higher rule of law indicates better governance which demonstrates an ability to implement change
Corruption	Better control of corruption indicates a greater likelihood of proper allocation of funds for adaptation
Education	Higher education indicates a higher skills base to change

Sources: Compiled from Knight (2011), Alberini *et al.* (2006) and Ullah (2005).

Maplecroft (2011) ranked 'the top 10 countries at 'extreme risk' from the impact of CC by using the CCVI in 2011 (pp. 2–3). The analysis measures vulnerability to CC by country on a relative basis. This provides insights into long-term strategic decision-making for corporates and policy-makers in the context of potential CC risk (Huddleston, 2010; Knight, 2011).

Today it is important to ask who emits more and suffers less. The level of discrepancies prevailing in the world system in GHG production by country as well as globally is shocking. The Western world – frontline campaigners for a green world – emits the most GHGs. However, China and India also emit alarmingly large amounts. As discussed above, CC is closely linked to energy production and consumption since fossil-fuel usage is the single biggest contributor to GHG emissions (Eckerberg and Forsberg, 1998; Birol, 2011). I often argue that a visible reluctance to protect the world could be explained by the Western-centric ideology that contends that there should be no concern about emitting GHGs because technology is there to protect us from environmental damage. Effective action against CC would reduce industrial productivity, which would in turn lead to job cuts. Therefore technology can face or control the damage that CC may cause (Anderson and Liefferink, 1997). Some critics and analysts (such as Dimitrov, 2010) also say that the failed Copenhagen summit 2009 was the corollary of such an approach.

4. Findings on climate change and economic effect correlates

GHGs are fundamentals to the world's energy system, which implies that they are somewhat benign (Doern and Gattinger, 2004). The production of CO_2 is intrinsic to fossil-fuel combustion. Specifically, thermal energy is generated by breaking the chemical bonds in the carbohydrates contained in oil, coal and natural gas, and oxidising the components to form CO_2 and H_2O. Cheap energy cannot be produced without CO_2 emissions. Similarly, methane (CH_4) emissions are necessary to prevent the build-up of hydrogen in anaerobic digestion and decomposition. Practically, this means beef, mutton, dairy and rice cannot be produced without methane emissions (Tol, 2009). One may ask why we can't stop emitting GHGs. This could put us into a dilemma between stopping GHG production and food production. The fact is that the emission of GHGs is essential for human survival, while being damaging at the same time. In most cases, GHGs come from energy use, which, in turn, is driven largely by economic growth. Therefore the minimisation of GHG emission is the best option.

The above paragraph tends to sum up the fact that, irrespective of the size, all manufacturing companies, farms and households have to emit some GHGs. This is because of the complex interdependences of the natural system, such as weather affecting agriculture, energy use, health and many other aspects of nature, which, in turn, affect everything that is part of nature. Hence the causes and consequences of CC are very diverse, and those in low-income countries, who contribute least to CC, as mention earlier, are most vulnerable to its effects. CC has appeared as the primary source of all externalities – larger, more complex and less certain than any other environmental problems. The sources of GHG emissions are more diffuse than any other environmental problem.

Why are GHGs so devastating to the world? The reasons are obvious and many. First, GHGs stay in the atmosphere for tens of thousands of years. Second, the quantities being emitted are enormous – for instance, in 2000, CO_2 emissions alone (and excluding land-use change) were 24 billion metric tons (Tol, 2009). More interesting and astonishing is that if all emissions were priced at the January 2009 value of €15/tCO_2 as applied in the Emissions Trading System of the European Union, CO_2 would be worth 1.5 per cent of world income (Tol, 2009; UN, 2012).

As considered earlier, is it, in fact, possible to measure the economic damage that CC can cause? The answer would obviously be no. However, the link between CC and the economic damage could be clarified. The increased costs of damage from storms, hurricanes, typhoons, floods, droughts, and heat waves counteract some early benefits of CC, and the cost could reach 0.5 to 1 per cent of world GDP per annum by the middle of the century. Some examples may dispel Eurocentric ideas that damage caused by climatological whims could be overhauled by the Western countries because they have the necessary resources. Flooding in the United Kingdom may cause losses worth 0.1–0.4 per cent of GDP once the increase in global average temperatures reaches 3 °C or 4 °C (Stern, 2007). The heat wave experienced in 2003 in Europe resulted in 35,000 deaths (although Larsen (2006) claimed the number was 52,000). Agricultural losses reaching $15 billion are directly linked to CC (Römisch, 2009).

'Climate change in India could cause a decline of around 20 million metric tons (25%) in rice production and over 30 million metric tons in wheat (30%) over 2000–2050' (Gerald *et al.*, 2010, p. 41). Clearly, this could create significant population pressure, causing domestic and international migration. For instance, millions of Indians live in areas that are vulnerable to flooding and water stress and, in particular, at least 12 million people in Mumbai alone are subject to flooding (IPCC, 2001).

There is no doubt that there will be an enormous impact of CC on the global economy (Daron, Johnson and Robinson, 2002). The tsunami of 2004, the most tragic and disastrous climatic catastrophe in recent history, has warned as well reminded us about possible losses that may reach an irreparable scale. According to the Asian Disaster Preparedness Centre (ADPC), the economic loss from the 2004 Indian Ocean earthquake and tsunami disaster was around US$10 billion. Three-quarters of the losses were incurred in four countries – namely, Indonesia, Thailand, Sri Lanka and India. Indonesia was the hardest hit in terms of human and physical damage

(ADPC, 2006). These losses were not only about lives but also about the destruction to residential and commercial buildings and infrastructure, and other healthcare facilities.

In those four countries, 'by 2100, the mean cost could reach 2.2 per cent of GDP each year if one considers market impact only, 5.7 per cent of GDP if non-market impacts related to health and ecosystems are included, and 6.7 per cent of GDP if catastrophic risks are also taken into account' (Brömmelhörster, 2009, p. 9).

Not only the loss of human life, homelessness and displacement of populations but also the macroeconomic impact of the disaster has been remarkable. The economic impact has become obvious and also that the poor would become poorer (Ullah, 2012a, 2012b). The large majority of the damage was along the west coast of Sumatra, Indonesia, and the loss was equivalent to approximately US$4.5 billion; and to a lesser extent the countries of India and Sri Lanka were affected (Table 3.2). The economy of the Maldives was severely hit and the loss there amounted to 45 per cent of GDP (Risk Management Solution (RMS), 2006).

4.1 The impact on population displacement

The discourse on whether CC is natural or anthropogenic lies in the following statement made by Zoe Knight (2011):

Even without climate change, soaring demand for natural resources on the back of demographic growth, economic expansion and the shift in the economic axis towards the emerging world is driving up commodity prices and intensifying resource risks, particularly at the intersection of energy, food and water. (p. 5)

A widespread misconception is that environmental change leads directly to migration. However, that is not the case all of the time. Rather, CC interacts with a range of economic, social and demographic factors (O'Connel, 2003).

Table 3.2 Economic loss caused by climate change

Country	Economic losses ($ million)	Insured losses ($ million)
Indonesia	4,500	500
Thailand	1,000	500
Sri Lanka	1,000	100
India	1,000	100
Maldives	500	50
Other	2,000	50
Total	10,000	1,300

Source: Canada Geological Survey (2006) and Ullah (2012a, 2012b).

This is especially the case when projecting the impact of CC. Population mobility is either a direct or an indirect response to environmental change and it can take many forms. Climate migration, then, may be viewed as a wide array of mobility types and not just displacement. It is seen as only one of the responses among many potential adaptation strategies that populations undertake. Urbanisation in Asia occupies an important space in the debate about CC since the urban population is growing dramatically with a strong coastal concentration. From about 24 per cent of the total in 1970 to 42 per cent in 2007, the population is likely to reach 50 per cent in the next decade (UN, 2012).

5. Discussions

It is relevant to bring minimalist and maximalist thoughts into the discussion in order to see how differently they view CC as a cause for the displacement decision. The migration-environment research literature tends to fall into two extreme categories – namely, (i) 'minimalists', who suggest that the environment is only a contextual factor in migration decisions, and (ii) 'maximalists or naturalists', who claim that the environment causes people to be forced to leave their homes. The maximalists and minimalists have missed out the economic factors. Irrespective of economic and habitat loss, everyone is either directly or indirectly affected. Thus Western centrism is getting weaker, though different states differ among themselves in managing the problem. Ultimately, CC governance and the problems of governance have become prominent in the global governance agenda. As mentioned earlier, minimisation of the impact of CC could be the best option in the present situation. Hence, climate governance merits space in the current discussion.

Climate governance represents significant challenges globally in terms of administrative and political systems of states through which they address this issue. Since the problem of CC is a global one, the UN system should have an active and significant role in addressing this problem. Multilateral environmental agreements (MEAs) are one way to address CC. One example of these is the UN Framework Convention on Climate Change (UNFCCC), 2007). This is a specific convention for dealing with the CC problem that was adopted at the Rio Earth Summit in 1992. Through this convention a legal structure has been formulated to deal with the matter in any future agreement. Also, it has determined the goals that need to be achieved in climate policy (Bernauer and Böhmelt, 2012).

In particular the UNFCCC is responsible for protection and ensuring that anthropogenic interference does not play a damaging role in the global environment. Most important is monitoring the effectiveness of regional and bilateral efforts that were supposed to reduce the volume of GHG emissions (Drexhage and International Institute for Sustainable Development, 2008; Crompton *et al.*, 2011). The UN Watercourses Convention has been drafted but has not yet entered into force regardless of the critical situation of water on the international political agenda. This is reminiscent of the slow response to attendant problems arising from CC (Gibson *et al.*, 2005). CC governance requires strategic capacity, which can be addressed through leadership, knowledge and provision of expert advice, 'defining the national interest and elaborating a strategic policy framework, and building organisations focused on a low carbon emission economy' (Meadowcroft, 2009, p.12). A need for an external entity that has an 'independent oversight body would oversee a range of national and private registries' (Drexhage and International Institute for Sustainable Development, 2008, p. 4). Far-reaching environmental, social and economic consequences, leading to political instability, increased income disparity, and the loss of biodiversity and habitats occurred, which negatively impacted the whole matter (ITTO-International Tropical Timber Organisation and FAO, 2009). Here is the need for a reform to the policies and legal frameworks (ITTO and FAO, 2009).

There is no denying that a global trend of migration is obviously triggered by CC. On one side there are people who move voluntarily in anticipation of environmental change. On the other side there are people who are forced to flee their homes as a result of environmental disasters (ADB, 2009a, 2009b). Thus this type of migration seems to directly result from CC and takes two major forms, which are 'migration associated with real or perceived direct environmental hazards, and migration associated with real or perceived reduced access and effective use of natural resources, including land, water, soil or biological resources' (ADB, 2009a, p. 9).

It is worth mentioning that CC policies can achieve their goals by being integrated into the development discourse or socioeconomic sectors, such as energy, transportation and industry. Integration implies an attempt to find synergies among different types of goal (Meadowcroft, 2009, p. 17). The mitigation of climate impact is a complex job and it has become more so due to the fact that it is a global feature and that the climate regime lacks centralised enforcement mechanisms (Geels, 2005; Bernauer and Böhmelt, 2012).

A few examples are worth mentioning. On a national level, joint commitments among local governments have emerged. The Nottingham Declaration, which was signed by 200 local authorities in the United Kingdom, is such an example, aiming at meeting specific climatic, culture and economic conditions (Morlot *et al.*, 2009). In order to tackle CC, the president provided a strategic direction for CC governance in the Philippines. He is responsible for directing and guiding all governmental agencies in order to work on their local development plans. Furthermore, these plans will be integrated into a regional form to provide regional development plans. This will require the regional offices of national governments, civil society organisations and provincial executives to be engaged.

The financial and investment elements are significant questions in the domain of CC governance. International efforts are not represented in the area of investment. These efforts are not institutionalised in order to 'foster the critically needed flows of clean energy investment in developing countries and for helping to ensure that they foster development' (Drexhage and International Institute for Sustainable Development, 2008, p. 4). The reason behind this is that the investment system is not unified but is scattered among many bilateral investment treaties. Essentially, the institutionalisation of CC within the domestic system of governmental organisations would 'effectively create 'champions' for mitigation and adaptation within governments of developing countries' (Drexhage and International Institute for Sustainable Development, 2008, p. 5).

Efforts of CC mitigation can be summarised into two major categories: economic and political factors. For the economic factors, economic growth has complicated effects on pollution. Yet, there is an argument for the benefit of this complex which is that bad environmental conditions usually lead to a better and clean environment. For example, economic growth is associated with updated technologies which are supposed to be environmentally cleaner according to the demand of the public for the sake of the environment (Bernauer and Böhmelt, 2012). In terms of the political system and how it can help in CC mitigation, political systems do have implications in this domain. For instance, democratic systems are excellent providers of good-quality environment-related products.

In an explanation of the measures that can be taken with regard to CC mitigation, these measures can include a requirement to incorporate CC impacts into national and regional planning processes, such as land-use and transport planning. Also, there is a need for periodic reports that highlight adaptation and anticipated long-range adaptation cost on the national and regional levels. In addition, these measures can embrace the establishment of regional and sector-based adaptation forums with key stakeholders to explore impacts; collaboration with the insurance industry to identify vulnerabilities and take remedial action; the integration of climate adaptation into agriculture and natural resource management plans; and the incorporation of adaptation issues into research-funding councils (Meadowcroft, 2009, p. 8). In addition to adaptation, governance of mitigation has specific requirements which are an understanding of the emissions source, cost-effective abatement potentials and policy approaches (Meadowcroft, 2009).

6. Conclusion

To sum up, because of the transnational nature of CC problems, a multi-level approach is needed. This should include states and international actors through the UN system, civil society, the private sector and stakeholders in order to represent economic, political and legal interests and benefits. Several megacities in South Asia, including Dhaka, Kolkata, Mumbai, Chennai and Karachi, are also at high risk of sea-level rise, more frequent cyclonic activity and greater saltwater intrusion. In Central Asia, widespread salinisation, inefficient water-management practices, land degradation, heat stress, desertification and increasing aridity are crucial issues that impede the social, cultural and economic well-being of local populations, exacerbating vulnerability to CC-induced effects in the local and regional environment.

This chapter concludes by posing some vital questions that merit answers because they have got a lot to do with the heart of the argument addressed here. This indicates that further in-depth research is necessary to ascertain the scale of potential damage which CC may cause to human life and livelihoods, and in human displacement. What particular environmental force has the potential to cause the most severe human displacement? Have researchers and policy-makers reached a consensus about the conceptualisation of the relationship between migration and the environment? What tools are available to measure the scale of migration caused by CC? How many more people are likely to migrate in the future? Since this chapter grasps the climatic and displacement issues in Asia, it would be an oversight if two giant economies of the region were not touched upon. China and India are among the largest GHG producers in the world. However, in the matter of compliance, China has recently declared its plan to reduce its GHG emissions while maintaining growth at the same pace as before. This means that it is aiming for energy-efficient production and introducing new technology into the production sector. Meanwhile, India has no such plans so far but the government has to comply with major accords and has agreed to reduce the emissions.

Abbreviations

ADPC	Asian Disaster Preparedness Centre
CC	climate change
CCVI	Climate Change Vulnerability Index
FAO	Food and Agriculture Organisation of the UN
GDP	gross domestic product
GHG	greenhouse gas
IPCC	Intergovernmental Panel on Climate Change
ITTO	International Tropical Timber Organisation
MEA	multilateral environmental agreements
UNESCO	UN Educational, Scientific and Cultural Organisation
UNFCCC	UN Framework Convention on Climate Change

References

ADB (2009a) *Asian Development Outlook. Rebalancing Asia's Growth* (Manila: Asian Development Bank), p. 23.
ADB (2009b) *Climate Change and Migration in Asia and the Pacific*, Executive summary (Manila: Asian Development Bank), pp. 1–44.

ADPC (2006) *Regional Analysis of Socio-Economic Impacts of the December 2004 Earth-quake and Indian Ocean Tsunami* (Bangkok: ADPC).

W.N. Adger (2006) 'Vulnerability', *Global Environmental Change*, 16, 3, 268–81.

A.C. Alberini, A. Chiabai and L. Muehlenbachs (2006) 'Using Expert Judgment to Assess Adaptive Capacity to Climate Change: Evidence from a Conjoint Choice Survey', *Global Environmental Change*, 16, 2, 123–44.

A. Ananthaswamy (2009) 'Sea Level Rise: It's Worse than We Thought', *New Scientist*, 2715, July 6, 2009.

M. Anderson and D. Liefferink (1997) *European Environmental Policy: The Pioneers* (Manchester: Manchester University Press).

D. Anthoff, David, C. Hepburn, and Tol, Richard S J. (2009) 'Equity Weighting and the Marginal Damage Costs of Climate Change', *Ecological Economics*, 68, 3, 836–849.

P. Baer and T. Athanasiou (2007) 'Frameworks and Proposals: A Brief Adequacy and Equity-Based Valuation of some Prominent Climate Policy. Climate Policy Framework and Proposals', *Global Issues* paper 30, Heinrich Boll Stiftung.

T. Bernauer and T. Böhmelt (2012) 'National Climate Policies in International Comparison: The Climate Change Cooperation Index', *Environmental Science and Policy*, 25, 196–206.

F. Birol (2011) *A Glimpse into the Energy Future* (London: Imperial College).

J. Brömmelhörster (2009) *The Economics of Climate Change in Southeast Asia: A Regional Review* (Manila: Asian Development Bank).

Canada Geological Survey (2006) *Managing Tsunami Risk in the Aftermath of the 2004 Indian Ocean Earthquake and Tsunami*, (Newark, CA: Risk Management Solution).

Christian Aid (2007) *Human Tide: The Real Migration Crisis: A Christian Aid Report, May 2007* (London: Christian Aid).

R.P. Crompton, R.A. Pielke Jr. and K.J. McAneney (2011) 'Emergence Timescales for Detection of Anthropogenic Climate Change in US Tropical Cyclone Loss Data', *Environmental Research Letters*, 6, 1–4.

S.H. Cullen and M. G. Sarah (2007) 'Trends and Triggers: Climate, Climate Change, and Civil Conflict in Sub-Saharan Africa', *Political Geography*, 26, 6, 695–694.

A. Daron, S. Johnson and J.A. Robinson (2001) 'The Colonial Origins of Comparative Development: An Empirical Investigation', *American Economic Review*, 91, 4, 1369–1401.

A. Daron, S. Johnson and J.A. Robinson (2002) 'Reversal of Fortune: Geography and Institutions in the Making of the Modern World Income Distribution', *Quarterly Journal of Economics*, 117, 4, 1231–1294.

R.S. Dimitrov (2010) 'Inside UN Climate Change Negotiations: The Copenhagen Conference', *Review of Policy Research*, 27, 6.

B. Doern and G. Gattinger (2004) *Power Switch: Energy Regulatory Governance in the Twenty First Century* (Toronto: University of Toronto Press).

J. Drexhage, International Institute for Sustainable Development (IISD) (2008) 'Global Environmental Governance (GEG)', Briefing Paper 2, Climate Change and Global Governance which we Ahead?, pp. 1–6.

P. Driessen and P. Glasbergen (eds.) (2002) *Green Society: The Paradigm Shift in Dutch Environmental Politics* (Dordrecht: Kluwer Academic).

A. Duit (2008) *The Ecological State: Gross National Patterns of Environmental Governance Regimes, EPIGOV Paper No. 39* (Berlin: Ecologic Institute for International and European Environmental Policy).

J. Ducanes and M. Abella (2009) *The Future of International Migration to OECD Countries: Regional Note – China and South East Asia* (Paris: Organisation for Economic Co-operation and Development).

K. Eckerberg (2000) 'Sweden: Progression Despite Recession', in W. M. Lafferty and J. Meadowcroft (eds.) *Implementing Sustainable Development Strategies and Initiatives in High Consumption Societies* (Oxford: Oxford University Press).

K. Eckerberg and B. Forsberg (1998) 'Implementing Agenda 21 in Local Government', *Local Government*, 3, 334–347.

J.J. Ewing (2012) *'Contextualising Climate as a Cause of Migration in Southeast Asia'*, *Chapter 2, p. 13. RSIS Monograph No. 24, Lorraine Elliott. S. Rajaratnam School of International Studies* (Singapore: Nanyang Technological University).

FAO (2009) *Nepal Forestry Outlook Study*, Working Paper No. APFSOS II/WP/2009/05 (Rome: Food and Agriculture Organisation of the UN).

F. Fung, A. Lopez and M. New (2011) 'Water Availability in +2°C and +4°C Worlds', *Phil. Trans. R. Soc. A*, 369, 67–84.

R. Garnaut (2008) *The Garnaut Climate Change Review* (Cambridge: Cambridge University Press).

F. Geels (2005) *Technological Transitions and System Innovations: A Co-evolutionary and Socio-technical Analysis* (Cheltenham: Edward Elgar).

C.N. Gerald, M.W. Rosegrant, A. Palazzo, I. Gray, C. Ingersoll, R. Robertson, S. Tokgoz, T. Zhu, T.B. Sulser, C. Ringler, S. Msangi and L. You (2010) *Food Security, Farming, and Climate Change to 2050: Scenarios, Results, Policy Options* (Washington D.C.: International Food Policy Research Institute).

R. Gibson, S. Hassan, S. Holtz, J. Tansey, and G. Whitelaw (2005) *Sustainability Assessment* (Oxford: Earthscan).

R. Heltberg (2007) 'Helping South Asia Cope Better with Natural Disasters: The Role of Social Protection', *Development Policy Review*, 25, 6, 681–698.

R. Heltberg and P.B. Siegel and S.L. Jorgensen (2009) 'Addressing Human Vulnerability to Climate Change: Toward a "No-Regrets" Approach', *Global Environmental Change*, 19, 89–99.

R. Heltberg and P.B. Siegel and S.L. Jorgensen (2010) 'Social Policies for Adaptation to Climate Change', in R. Mearns and A. Norton (eds.) *Social Dimensions of Climate Change: Equity and Vulnerability in a Warming World* (Washington D. C.: World Bank).

T.W. Hertel, M.B. Burke and D.B. Lobell (2010) 'The Poverty Implications of Climate-Induced Crop Yield Changes by 2030', GTAP Working Paper No. 59. Global Trade Analysis Project.

M. Huddleston (2010) 'Managing Weather Risk in a Changing Climate: Opportunities from the Developing Science', The Met Office, March.

R. Holzmann and S. Jorgensen (2000) 'A New Conceptual Framework for Social Risk Management and Beyond', *Social Protection Discussion Paper 9926* (Washington, D.C.: The World Bank, Intergovernmental Panel on Climate Change.

G. Hugo (2008) *Migration, Development and Environment. International Organisation for Migration*, IOM Migration Research Series. No. 35 (Geneva: International Organisation for Migration).

IPCC (2007a) *Fourth Assessment Report* (Climate Change 2007) (Geneva: Intergovernmental Panel on Climate Change).

IPCC (2007b) *Climate Change 2007: Synthesis Report* (Geneva: Intergovernmental Panel on Climate Change).

IPCC (2007c) *Climate Change 2007: Contribution of Working Groups I, II, and III to the Fourth Assessment Report of the Intergovernmental Panel on Climate Change* (Cambridge: Cambridge University Press).

IPCC (2012) Field, C.B., V. Barros, T.F. Stocker, D. Qin, D.J. Dokken, K.L. Ebi, M.D. Mastrandrea, K.J. Mach, G.K. Plattner, S.K. Allen, M. Tignor, and P.M. Midgle

"Managing the Risk of Extreme Events and Disasters to Advance Climate Change Adaptation" (Cambridge: Cambridge University Press).

ITTO and FAO (2009) 'Forest Governance and Climate-Change Mitigation' in T. Bernauer and L. M. Schaffer (eds.) *Climate Change Governance* (Zurich: Institute of Environmental Decisions), pp. 1–28.

IPCC (2011) *Climate Change 2001: Synthesis Report* (Geneva: IPCC).

S. Idean (2008) 'From Climate Change to Conflict? No Consensus yet', *Journal of Peace Research*, 45, 3, 315–326.

Z. Knight (2011) 'Scoring Climate Change Risk: Which Countries are most Vulnerable?' HSBC Global Research.

J. Larsen (2006) 'Setting the record straight: More than 52,000 Europeans died from heat in summer 2003', *Report* (Washington D.C.: The Earth Policy Institute).

Maplecroft (2011) Big economies of the future – Bangladesh, India, Philippines, Vietnam and Pakistan – most at risk from climate change. Climate Change Vulnerability map (United Kingdom: St Stephen's).

J.C. Morlot, C.L. Kamal, M.G. Donovan, I. Chochran, A. Robert and P.J. Teasdale (2009) 'Cities, Climate Change and Multilevel Governance', *OECD*, pp. 1–125.

J. Meadowcroft (2009) *Climate Change Governance* (Washington D.C.: The World Bank, Development Economics, World Development Report Team), pp. 1–40.

N. Myers (2002) 'Environmental Refugees: A Growing Phenomenon of the 21st Century', *Philosophical Transactions of the Royal Society B*, 357, pp. 609–613.

N. Myers (2005) 'Environmental Refugees: An Emergent Security Issue', *13th Economic Forum*, Prague, 23–27 May.

R.J. Nicholls, P.P. Wong, V.R. Burkett, J.O. Codignotto, J.E. Hay, R.F. McLean, S. Ragoonaden and C.D. Woodroffe (2007) 'Coastal Systems and Low-lying Areas. Climate Change 2007: Impacts, Adaptation and Vulnerability. Contribution of Working Group II to the Fourth Assessment Report of the Intergovernmental Panel on Climate Change', in M.L. Parry, O.F. Canziani, J.P. Palutikof, P.J. van der Linden and C.E. Hanson (eds.) (Cambridge: Cambridge University Press), pp. 315–356.

M.J. O'Connel (2003) *Carbon Constraints in the Building and Construction Industry: Challenges and Opportunity*, BRANZ Issue Paper No. 2 (Porirusa, New Zealand: BRANZ).

R.Z. Poore, R.S. Jr. Williams and C. Tracey, Christopher (2000) 'Sea Level and Climate: U.S. Geological Survey Fact Sheet 002–00', http://pubs.usgs.gov/fs/fs2-00/ home page, date accessed 19 October 2012.

R. Reuveny (2007), 'Climate Change-Induced Migration and Violent Conflict', *Political Geography*, 26, 656–673.

RMS (2006) *Managing Tsunami Risk in the Aftermath of the 2004 Indian Ocean Earthquake & Tsunami* (CA: Risk Management Solution Inc.).

R. Römisch (2009) 'Regional Challenges in The Perspective of 2020 and Regional Disparities and Future Challenges', *Background Paper on Climate Change* (EU: European Commission).

N. Stern (2007) *The Economics Of Climate Change: The Stern Review* (Cambridge: Cambridge University Press).

S.J.R. Tol (2009) 'The Economic Effects of Climate Change', *Journal of Economic Perspectives*, 23, 2, 29–51, Spring.

A.K.M.A. Ullah (2005) 'Poverty Alleviation in Bangladesh: "Does Good Governance Matter?"' In J. Jabes (ed.) *The Role of Public Administration in Alleviating Poverty and Improving Governance* (The Philippines: The Network of Asia-Pacific Schools and Institutes of Public Administration and Governance (NAPSIPAG) (Manila: Asian Development Bank)).

A.K.M.A. Ullah (2010) *Rationalising Migration Decision: Labour Migrants in South East and East Asia* (Aldershot: Ashgate).

A.K.M.A. Ullah (2012a) 'Climate Change and Climate Refugee in Egypt: An Overview from Policy Perspectives', *TMC Academic Journal*, 7, 1, 56–70.

A.K.M.A. Ullah (2012b) *Displaced, Disabled and Disturbed: Narratives of Trauma and Resilience among Acehnese Survivors of the 2004 Tsunami*, NTS-Asia Research Paper No. 7 (Singapore: RSIS Centre for Non-Traditional Security (NTS) Studies for NTS-Asia).

A.K.M.A. Ullah and M.A. Hossain (2011) 'Gendering Cross-Border Networks in Greater Mekong Sub-region: Drawing Invisible Routes to Thailand', *Austrian Journal of South-East Asian Studies*, 4, 2, 273–289.

UN (2012) *Sustainable Poverty Reduction and Natural Disaster Risk Management* (Vietnam: United Nations).

UNESCO (2009) *Water in a Changing World', The United Nations World Water Development Report 3* (Paris: UNESCO).

UNFCCC (2007) *Investment and Financial Flows to Address the Climate Change*, Background paper (Bonn: UNFCCC Secretariat).

US Department of Transportation (2010) *Transportation's Role in Reducing U.S. Greenhouse Gas Emissions*, Vol. 1 Synthesis Report 1 (Washington D.C.: US Department of Transportation).

4
Disaster Communication in Mitigating Climate Change in Sri Lanka: Problems and Prospects

R. Lalitha S. Fernando

1. Introduction

Climate change (CC) as a natural disaster is a serious threat to humanity. As Sri Lanka is situated in the Indian Ocean, and the country as a whole is now more vulnerable to the effects of CC than ever before. A serious disruption was experienced by the people in Sri Lanka during the tsunami on 26 December 2004, which is considered to be the result of CC. It was an unforgettable incidence for the country's citizens. The impact on people's lives and the economy has not been accurately estimated yet. Table 4.1 shows the number of people who were affected by various types of disaster during the period of 1974–2004 in Sri Lanka.

The table shows that Sri Lanka is highly vulnerable to the effects of various types of CC. Though any natural disasters cannot be prevented, many human lives could have been saved if those affected had been warned in good time. According to the table, during the period of 1974–2004, about 2,964,655 people were affected due to flooding, which is now a common phenomenon in the country. The tsunami was considered to be the worst incident in living memory. Table 4.2 shows the impact of the tsunami.

Most effects of any disaster could have been minimised if the relevant authorities had taken proactive measures. Seven years after that painful and unforgettable experience, another catastrophe – a cyclone – occurred on 25 November 2011, resulting in considerable damage to people's lives and property, and to the economy. The cyclone, which some called 'squall line' (a condition of rain and wind), was reported in the southern part of the country. More than 30 fishermen died and the same number of people went missing. People in the south were also severely affected by this event. The stormy winds that hit the Matara and Galle districts in the Southern Province in Sri Lanka resulted in the loss of several lives, destroyed properties and

Table 4.1 The number of people who were affected by various types of disasters during the period of 1974–2004 in Sri Lanka

Type of disaster	Affected people (number)
Flood	2,964,655
Tsunami	1,009,474
Landslide	46,719
Drought	2,072,512
Cyclone	303,001

Source: Disaster Management Centre of Sri Lanka (2005), cited by Imesha (2011, p. 25).

Table 4.2 The impact of the tsunami in 2004

Types of impact	Number of cases
Deaths	30,196
Displaced persons	838,388
Injured persons	15,683
Missing persons	3,792

Source: District secretariats (2005), cited by Imesha (2011, p. 27).

Table 4.3 The impact of the cyclone of 25 November 2011

Types of damage	Number of cases
Affected families	8,854
Affected people	34,596
Deaths	29
Injuries	24
Missing people	14
Houses damaged (fully)	866
Houses damaged (partially)	8,889

Source: Based on the Situation Report on 5 December 2011 by the Disaster Management Centre of Sri Lanka (2011), retrieved from www.dmc.lk (accessed on 30 December 2011).

caused problems to fishermen. The impact on the economy has yet to be assessed. Table 4.3 shows the impact of the incident in the affected areas.

It is evident that the policies on disaster management in Sri Lanka were not effective. A lack of communication is considered to be one of the main obstacles in disaster management in the country. The reported damage could have been reduced if there had been an effective communication system.

A timely weather forecast could have prevented the fishermen from venturing to the sea ahead of the bad weather. After a disaster the relevant authorities in disaster management normally undertake hasty rehabilitative and recovery activities in response to the incident. Despite considerable efforts to design and implement a policy on CC mitigation in Sri Lanka, the attention that has so far been given to disaster communication seems to have been inadequate.

2. Research questions

Without communication, no disaster management is possible. Indeed, a lack of communication is one of the main constraints in mitigating CC in Sri Lanka. Disaster communication is made simple with information and communications technology (ICT). However, despite the adoption of e-government in Sri Lanka several years ago, the relevant authorities have not used this concept in a sensible and appropriate manner. This chapter identifies the issue of disaster communication as a barrier which has to be given the strongest consideration in any form of disaster management in Sri Lanka.

3. Research methods

This chapter reviews the literature on disaster management, CC and disaster communication to investigate the problem. A recent incident of CC was taken as a case study to illustrate the severity of the problem of disaster communication in Sri Lanka. The study is mainly based on secondary data, and a descriptive method was used to analyse and present the data.

4. Mitigating climate change

CC is regarded as unexpected changes in weather patterns over long periods of time. The United States Environmental Protection Agency (2009, cited in Baba, 2010) defines CC as any distinct changing measures of climate lasting for a long period of time. Increases in unexpected rainfall, storms, droughts, flooding, land slides, temperature and sea level are major effects of CC, and such an effect is considered to be a disaster if it creates a sudden or great misfortune in a community. Disasters happen due to natural or man-made causes. As CC is a natural disaster which cannot be prevented, any possible damage could be minimised through an effective disaster-management strategy. Some disasters occur suddenly while others can be identified before they happen. The Asian Development Bank (USAID Asia, 2010) defines adaptation in this context as adjustments to reduce costs and vulnerabilities based as a result of CC impacts. Adaptation activities are related to actions aimed at providing protection from CC impacts. There are two main methods for tackling CC – mitigation and adaptation. 'Mitigation aims to reduce greenhouse gas emissions, with the goal of slowing or preventing climate change, whereas adaptation is the act of reducing vulnerability to the impacts of climate change' (Sanderson and Islam, 2007, cited in Mair, 2011, p. 246). As disasters cause serious disruption to human life (loss of lives, illness or injury), and severe property loss and damage to the environment, the government's intervention is essential. As Sri Lanka is highly vulnerable to the impact of CC, the potential damage caused by any disaster has to be minimised and prevented with an effective disaster-management system.

In disaster management, arrangements have to be made to manage the potential adverse effects of an event in terms of mitigating, preventing, preparing for, responding to and recovering from a disaster. Various activities have to be undertaken, such as facing the situation successfully, distributing necessary goods and services, rehabilitating, rebuilding and reconstructing to minimise the effects of the particular disaster. Communication is essential in implementing each activity involved in disaster management. Without communication, it is impossible to achieve a better disaster management system. Thus communication is vital in managing any disaster.

5. The role of ICT in disaster communication

Communication is a process of passing information and understanding from one person to another. It involves transmitting and sharing ideas, opinions, facts and information in a manner that is perceived and understood by the receiver of the communication. Effective communication is essential for disaster management to perform its functions successfully. Many operations may fail due to incomplete communication, misunderstood messages and unclear instructions. Communication must be interpreted and understood in the manner intended by the sender. In order to have effective communication, several guidelines need to be followed. For example, the ideas and message should be clear, brief and precise. The message must be sent within an appropriate time, should be passed through the proper communication channels and it should be comprehensive. Barriers such as noise, poor timing, incomplete and inadequate or unclear information, physical distractions, information overload, and network breakdown can affect communication. Both traditional and modern methods are available in this regard.

In disaster management, communication means informing citizens about potential disasters. People have to be informed before any disaster, during the event and also afterwards. In disaster communication, e-government plays a vital role. 'E-government is defined as governments' efforts to provide citizens with the information and services they need, using a range of information and communication technologies' (Robins, 2003, cited in Mofleh and Wanous, 2008, p. 1). The use of ICT in public services delivery

is also known as digital government, online government or virtual government, where public services are delivered through government institutes via
telephone, mobile phone, SMS (Short Message Service), fax, television- and
radio-based government services, e-mail or websites. ICT can play a crucial
role in disaster-management activities in terms of preparedness, mitigation,
relief and recovery. Similar to other public services, clever use of ICT can
make disaster management more effective, especially in terms of prevention
and mitigation. Correct, timely and important information is a crucial factor
to any successful disaster-management programme and geographic information system (GIS) could provide this kind of information – for example, in
identifying the potential risk so as to reduce its impact, and also in identifying appropriate measures for disaster prevention, mitigation or preparation.
Thus GIS could be used to identify the potential disaster before the event
and also to analyse the consequences afterwards.

In a report by the United Nations Development Programme (UNDP) –
Asia-Pacific Development Information Programme (2007) entitled *ICT for
Disaster Management in 2007*, Krasae Chanawong emphasises the importance of remote sensing for early warning which is made possible by various
technologies, including telecommunication satellites, radar, telemetry and
meteorology. The report explains that ICT encompasses both traditional
media (radio, television) and new media (cell broadcasting, internet, satellite
radio), all of which can play a major role in educating the public about
the risks of a potential or impending disaster. Before a disaster, ICT could
be used to disseminate information about the event and its potential danger, and strategies and activities which will be implemented to mitigate
its potential impact. Communication is essential before, during and after
a disaster because it creates an emergency situation. The relevant authorities need to take responsibility for sending the right message (in terms of
content, length and format); monitor its delivery and response; and ensure
that the process is initiated and suspended at the right times (Page, 2006).
The report describes various strategies for addressing communication challenges: (i) the use of automated notification technology which can rapidly
distribute information to a large number of people, (ii) the implementation
of training programmes, feedback systems and regular testing that could be
conducted so as to reduce human errors (e.g. sending incorrect messages or
failing to notify the right parties), and (iii) using multiple modes of contact
through emails, telephone (mobile and landline), and the media (television
channels, radio broadcasting, newspapers) that can be used to reach citizens
in the face of communication failures and obstacles. In this process, at the
request of the relevant authorities, both public and private sector Medias
could meet their social responsibility by disseminating disaster information
and ensuring appropriate warnings at the right time.

If citizens are well informed about the disaster, they are less likely to
fall victim to it. For example, if people had been informed about the

Table 4.4 Time gap in the occurrence of the tsunami on 26 December 2004

Time	Incident	Number of deaths recorded
8.27 a.m.	A tsunami occurred in Kalmunei (Ampara District/northern part)	10,436
8.40 a.m.	Baticcaloa (north-eastern part)	2,254
8.55 a.m.	Mulathiw (northern part)	2,000
8.55 a.m.	Trincomalee	947
9.15 a.m.	Valvatithurei (Jaffna District)	2,640
9.15 a.m.	Hambanthota (southern part)	4,500
9.15 a.m.	Matara	1,061
9.20 a.m.	Galle	3,724
9.20 a.m.	Kalutara	213
9.34 a.m.	First message was broadcast as on the condition of the tsunami in Sri Lanka.	

Source: Disaster Management Centre of Sri Lanka (2005), cited by Imesha (2011, pp. 47–48).

aforementioned tsunami, thousands of civilian deaths could have been avoided. In this regard, relatively inexpensive forms of communication could have been used, such as televisions, mobile phones and landlines (through SMS). Table 4.4 shows the importance of disaster communication related to the tsunami in Sri Lanka.

It is evident that there was about 45–60 minutes between the northern and southern parts of the country being affected by the tsunami, so with an effective system of disaster communication, thousands of civilian lives could have been saved in the south. It is important to note that the cost of disaster communication as a preventive measure may not be that expensive when compared with the implementation of rehabilitative activities in reaction to a disaster.

6. Policy initiatives on disaster management in Sri Lanka

Until the early 1990s, little attention was given to formulating policies for disaster management in Sri Lanka. In 1991, the government appointed a subcommittee of the cabinet to formulate a national policy for disaster management. Even though no policy existed, a special institution called the National Disaster Management Centre (NDMC) was established for the first time in 1996 under the Ministry of Social Services (MSS). These two organisations, the NDMC and the MSS, have since been involved in disaster management. The NDMC has been involved in policy planning and coordinating organisations that have taken part in disaster-management activities. The MSS has been involved in post-disaster activities with the assistance of district and divisional secretariats. As there was no systematic legal system for activities related to pre-disaster and post-disaster situations, the rehabilitative, reconstructive and other related actions were undertaken under the Emergency Law of Sri Lanka. In 2001, a policy document on disaster management was forwarded to the parliament for necessary approval. The National Environmental Action Plan (1998–2001) established by the National Council for Sustainable Development was also related to disaster-management activities in Sri Lanka. However, after the tsunami, the government of Sri Lanka has prioritised disaster management. After the tsunami in 2004, two important acts were passed: the Sri Lankan Disaster Management Act No. 13 of 2005 and the Sri Lanka Tsunami (Special) Act of No. 16 of 2005. Under the Disaster Management Act, several initiatives have been introduced: a special organisational framework has been set up to implement disaster-management policies; the process of disaster management has been specified; and the National Advisory Council for Disaster Management has been established. The president of Sri Lanka is the chair of this council. Other members are the prime minister, the leader of the opposite political parties and other ministers (e.g. the minister of disaster management). Several organisations work under the Ministry of Disaster Management, which is the main organisation responsible for policy-making in disaster management.

The Disaster Management Centre (DMC) has been established in accordance with the National Council for Disaster Management under the Disaster Management Act. It functions under the Ministry of Disaster Management and Human Rights. The organisational framework for disaster management in Sri Lanka is depicted in Figure 4.1.

In 2008, a ten-year environmental plan named 'Haritha Lanka' was designed and, under this, an adaptation action plan and mitigation plan were designed. However, these plans were not implemented due to the change of the ministerial portfolio. In 2010, the National Climate Change Adaptation Strategy (NCCAS) of Sri Lanka for 2011–2016 was introduced. Whatever policies and plans were initiated by the government of Sri Lanka for mitigating adverse impacts of CC, and despite the president of Sri Lanka being the chair of the established National Advisory Council for Disaster Management, all of the efforts made with regard to disaster management have not been successful. The following case provides some evidence in this regard.

7. Issues of disaster communication are part of the problem

After the disaster of December 2011, citizens and the media, including the Minister of Disaster Management in Sri Lanka, criticised officials of the Meteorological Department for their poor weather forecasting. The minister requested a report from the Meteorological Department within three days of the incident because of the department's failure to forecast the recent catastrophic weather: 'Disaster Management Minister called for a report from the

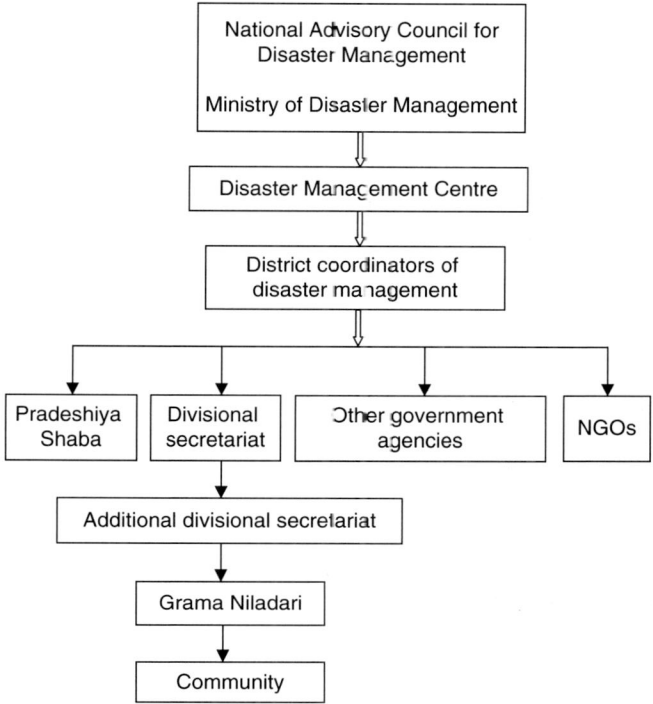

Figure 4.1 Organisational framework for the Disaster Management Centre of Sri Lanka
Source: Disaster Management Centre of Sri Lanka (2005), cited by Imesha (2011, p. 43).

Metrological Department to ascertain whether there was any lapse behind the Department' (Ranil, 2011, p.14). The Meteorological Department is often criticised by the media and the citizens of Sri Lanka due to poor weather forecasting. It is clear that disaster represented huge losses to the country. The Meteorological Department has taken responsibility for the failure in disaster communication related to this case. The chief of the department stated: 'Actually, it was the Weather Forecasting Division which should be responsible, but as the Director General of the Department, I have to take that responsibility' (Ranil, 2011, p. 14). It was also reported that the website of the Meteorological Department had not been updated since October 2011. This indicates the lack of commitment and responsibility of its officials. According to the Meteorological Department, normally it released its weather forecasts to the media and the DMC. The reports are sent to the respective district secretariats. However, the department had not taken the measures necessary to mitigate the situation.

A lack of resources is part of the problem in disaster communication. A shortage of competent people is another part of the problem, which is related to the inefficiency of disaster communication in the Meteorological Department. The department chief said: 'If we had Doppler Weather Radar, we could have alerted people one hour before the catastrophic struck' (Ranil, 2011, p. 14). Further, the head of the department urged the necessity of building the capacity to respond to such a situation. With regard to the human resources problem, the Meteorological Department says that it has only one duty forecaster at a time. The chief stated: 'One person cannot handle all the things. It is meteorologist who had to analyse the data, look into the charts, and forecast. It has to be done after processing and analysing the data that is very dedicated things' (Ranil, 2011, p. 14). At that interview the chief revealed that at the time of the disaster there were only 17 scientists in the department and six more vacancies needed to be filled. The head compared the situation with other countries and said that 'certain countries have different sections for cloud picture analysis, Doppler radar, optional data and upper data radar. So all assimilations come to the chief forecaster and he makes the forecast, but here at the Meteorological Department, we do not have people to deploy for different sections' (Ranil, 2011, p. 14). Thus the lack of manpower in the department is another part of the problem. However, despite the limited resources, disseminating information about the disaster could have been done to some extent via the relevant authorities with the assistance of the international agencies. Hence the officials' commitment and work motivation, and their proactive behaviour, are also problematic in this respect. The most important factor seems to be the ignorance and the lack of commitment of several officers in the department.

8. What can be done to improve disaster communication?

Establishing an immediate response system for natural disasters has become a vital strategy in mitigating the damage of such disasters. Effective communication is critical to ensure that the community can respond quickly, efficiently and effectively. Both traditional and modern methods of communication are equally important. However, more important is the use of modern communication methods, where ICT plays a significant role. Satellite phones, hand-held radios and mobile phones could also be used. The advancement of communication technologies also means that satellite transmission, wireless networks, the internet and other cost-effective communication methods are available (Yap, 2011). Thus the role of ICT in disaster management is of utmost importance. In this area the GIS could be used for highlighting risk areas, vulnerabilities and the potentially affected population. Electronic and printed media could also be effectively used as complementary methods in disaster communication. Both the public and private sector media, especially radio and television organisations, have a key role in disseminating disaster-related educational messages. These organisations have a responsibility to ensure that the warnings and advice about disasters are timely and appropriate at the request of the relevant authorities. Evidence shows the lack of the public's awareness and education about CC despite the availability of various protective measures. The Global System for Mobile Communication (GSM)-based cell broadcasting (CB) is one of the newest trends in telecommunications and is ideal for natural disaster responding. In a report by the UNDP, Chanuka Wattagama (2007) describes several advantages of CB systems. He further mentions that most of today's wireless systems support a feature called cell broadcasting, and a public warning message in text can be sent to the screens of all mobile devices with such capability in any group of cells of any size, ranging from one single cell (about 8 km across) to the whole country if necessary. Thus the GSM is capable of delivering information promptly to all mobile phones in a specific network in a predetermined geographical area. The location-based identification of mobile phones could be used as the basis of disaster communication. 'GSM is also presented as being able to make people to communicate with each other, on real time basis, saving time and money, among other conveniences. Facilitating access to up-to-date information to support real time decisions increases efficiency in environmental monitoring, disaster control and emergency management' (Elegbeleye, 2005, p. 197). Thus the GSM-based CB and SMS capabilities have enormous potential to overcome communication barriers in disaster management, and clever use of ICT could be considered as a cost-effective means of disaster communication.

9. Conclusion and policy implication

The development process today cannot be preceded in the absence of an effective weather forecasting system which monitors the weather patterns of the country by enabling citizens to overcome the problems of disasters. At the stages of development policy planning, the effects of CC have to be taken into consideration as the natural environment and development are closely interdependent. The Meteorology Department and other relevant institutions in Sri Lanka have an increasingly prominent and vital role in communicating disaster information to minimise the possible impact of CC. In this process the concept of e-government with the use of GIS could play a significant role. The electronic tools, mobile phones, telephones, emails, newspapers, radio, television, loudspeakers, alarms and so on could also be used to ensure the effectiveness of the implemented disaster communication. The use of mobile phones with SMS, particularly the GSM, has a vital role in this regard. An improved early-warning system could be used in disaster management as an effective mechanism to minimise potential damage resulting from a disaster. The authorities that are responsible for disaster management are required to cooperate and coordinate in channelling information across reliable communication systems for a better response before, during and after any disaster.

The Meteorological Department has a vital role in disseminating weather forecasts but other institutions, especially the DMC in Sri Lanka and public and private institutions, including the media, could also play a crucial role in communicating disaster information. In order to respond to any disastrous situation, cooperative and coordinative efforts of the government, the private and non-governmental sectors, the media and other organisations that are responsible for disaster management are required. However, the efficiency and effectiveness of any attempt at disaster management depend on the availability of adequate resources, competent and committed people, and the relevant authorisation. Civil society with the educated community should also participate in this process. In this context, the e-government could play a dominant role.

Beyond these technologies and strategies, individual creativity must be brought to bear in order to establish communication systems. The relevant authorities should play an increasingly prominent and vital role in communicating disaster information to minimise the possible impact of CC. All of these strategies could be considered as short-term measures in managing any disaster. The Department of Economic and Social Council of the United Nations (2007) emphasised the importance of linking sustainable development policies and measures with action on CC. Thus the concept of sustainable development has to be practised to the maximum extent as a proactive development strategy in the long term to manage any natural disaster.

Abbreviations

CB	cell broadcasting
CC	climate change
DMC	Disaster Management Centre
GIS	geographic information system
GSM	Global System for Mobile Communication
ICT	information and communications technology
MSS	Ministry of Social Services (Sri Lanka)
NCCAS	National Climate Change Adaptation Strategy
NDMC	National Disaster Management Centre
SMS	Short Message Service
UNDP	The United Nations Development Programme

References

N. Baba (2010) 'Sinking the Pearl of the Indian Ocean: Climate Change in Sri Lanka', *Global Majority E-Journal*, 1, 1 (June 2010), 4–16.

Climate Change Secretariat, Ministry of Environment and Asian Development Bank (2010) *National Climate Change Adaptation Strategy for Sri Lanka: 2011–2016* (Sri Lanka: Climate Change Secretariat, Ministry of Environment).

Department of Economic and Social Council, United Nations (2007) *Achieving Sustainable Development and Promoting Development Cooperation* (New York: United Nations).

Disaster Management Centre of Sri Lanka (2011) *Situation Report, November, 2011* (Colombo: Disaster Management Centre, Ministry of Disaster Management, Sri Lanka), http:// www.dmc.gov.lk, date accessed 12 January 2013.

O.S. Elegbeleye (2005) 'Prevalent Use of Global System of Mobile Phone (GSM) for Communication in Nigeria: A Breakthrough in Interactional Enhancement or a Drawback?', *Nordic Journal of African Studies*, 14, 2, 193–207.

D. Imesha (2011) *Disaster Communication: Theory and Practice, S. Godage and Brothers* (Colombo, Sri Lanka: Godage and Brothers).

J. Mair (2011) 'Events and Climate Change: An Australian Perspective', *International Journal of Event and Festival Management*, 2, 3, 245–253.

S.I. Moflehand and M. Wanous (2008) 'Understanding Factors Influencing Citizens' Adoption of e-Government Services in the Developing World: Jordan as a Case Study', *Journal of Emerging Trends in Computing and Information Sciences*, 2, 5, 1–11.

D. Page (2006) Emergency Communications and Disaster Response, http:www.crisissimulations.com, date accessed 20 January 2012.

W. Ranil (2011) Interview with the Metrological Department Chief, *Sunday Observer News Paper*, Lake House Publications in Sri Lanka, 4 December 2011.

United Nations Development Programme – Asia-Pacific Development Information Programme (2007) *ICT for Disaster Management in 2007* (Bangkok and Incheon: United Nations Development Programme – Asia-Pacific Development Information Programme).

USAID Asia (2010) *Asia Pacific Regional Climate Change Adaptation Assessment. Final Report: Findings and Recommendations* (Washington D. C.: International Resources Group).

C. Wattagama (2007) ICT for Disaster Management, the United Nations Development Programme – Asia-Pacific Development Information Programme (UNDP-APDIP) and Asian and Pacific Training Centre for Information and Communication Technology for Development (APCICT), http://www.apdip.net, date accessed on 15 January 2012.

N.T. Yap (2011) *Disaster Management, Developing Country Communities and Climate Change: The Role of ICTs* (Manchester, UK: Centre for Development Informatics Institute for Development Policy and Management, SED).

5
Climate Change and Global Environmental Governance: The Asian Experience

Gamini Herath

1. Introduction

Climate change (CC) is one of the most critical anthropogenic environmental impacts that has been experienced globally in recent decades. It can profoundly alter our economic future and livelihood. The Intergovernmental Panel on Climate Change (IPCC) has confirmed that the earth's climate is getting warmer, with the likely increase in temperature ranging from 1.1 °C to 6.4 °C and the likely increase in sea level ranging from 0.18 m to 0.59 m (IPCC, 1995). CC will have dire consequences for developing countries, exacerbating poverty, disease and instability.

According to the Department of Hydrology and Meteorology, the average temperature in Nepal is increasing at a rate of approximately 0.06 °C per year (Shrestha *et al.*, 1999). It is estimated that over the next 30–50 years the temperature may rise by about 2.6 °C in Malaysia. Warming in the country since the 1970s has been three times as rapid as for the preceding 100 years. The 1990s saw the warmest decade in the 142 years since temperature records began. A rise in sea level can affect food security for people located in coastal areas due to a rise in average temperatures (Raman, 2009).

The emission of greenhouse gases (GHGs) is the main reason for CC. Asia contributed 31 per cent of global emissions in 2006, and this is expected to rise to 42.1 per cent of global emissions in 2030 (Shui and Harriss, 2006). Much of the emissions have been due to forest clearing in Southeast Asian countries, which also destroyed considerable biodiversity and ecosystem services. Globalisation since the 1980s and rapid industrialisation have increased the emission of GHGs, thereby exacerbating CC. China's CO_2 emissions increased from 1,460 million tons in 1980 to 6,499 million tons in 2007, becoming the world's biggest emitter. The International Energy Agency estimates that China's CO_2 emissions will continue to increase to 11.4 gigatons in 2030 with business as usual (Shui and Harriss, 2006).

CC effects are most significant in tropical and subtropical regions which are poor and where governance capacity is inadequate (Schipper and Pelling, 2006). It can seriously affect developing countries in Asia where populations are most vulnerable and least capable of easily adapting to CC (Beg et al., 2002). The distributional impacts of such policies are very important because of the limited ability to adapt or to mitigate (Beg et al., 2002). Intergenerational equity is an important issue in CC because GHG emissions at one time can exert a considerable influence on events some decades later. Global CC will affect resource availability, which can reduce human development and growth.

Varying policies have been adopted in different countries to mitigate CC. In developing countries in Asia, however, the capacity for governance may be limited. CC transcends national boundaries, and hence global environmental governance mechanisms that go beyond state-led treaties have emerged (Biermann and Pattberg, 2008). Governments are equal partners here along with non-governmental organisations (NGOs), activist groups, citizens and other entities. Opportunities may exist in developing countries to effectively use global environmental governance (GEG) regimes to assist in the design of integrated responses to CC. How GEG can build cost-effective strategies and institutional capacity in developing countries to respond to CC needs to be critically evaluated.

The aim of this chapter is to critically review relevant published literature in order to synthesise CC challenges for Asia. The specific aims are to:

- review trends in CC and impacts in Asia;
- examine the adaptive and mitigation measures taken in Asian countries;
- critically evaluate challenges in implementation of selected global environmental governance initiatives;
- identify the implications of GEG for CC mitigation in Asia.

The methodology involves a review of the extant literature and synthesis of a narrative that reflect the achievements of GEG to date. The review covers literature up to 2010 to provide insights into the most recent developments of CC issues. The focus is mostly on the global governance of CC and why some of the major initiatives failed to have a significant effect on it.

2. Global environmental governance

GEG is a complex concept which refers to the nature and distribution of the institutionalised capacity to make decisions to manage global environmental problems such as CC. The 1972 United Nations (UN) Conference on the Human Environment in Stockholm was the precursor to GEG regimes of the 1980s and 1990s. GEG extends the power of states in the global system. The reason for an increase in GHG is the inadequacy of political responses to global environmental problems. GEG shifts the location of authority in political, economic and social realms, indicating a move away from the state-centric view of governance. The Commission on Global Governance (1995) defines governance as formal institutions and regimes empowered to enforce compliance as well as the informal arrangements that people or institutions have agreed or perceive to be in their interest (Commission on Global Governance, 1995).

French analyst Smouts (1998, p. 88) argued that global governance is not an 'analytical reflection on the present international system [but a] standard setting reflection for building a better world'. The aim of global governance is to ensure effective responses to environmental issues that transcend the capacities of the nation state to address individually. It is a pluralistic concept of the global system distributing power away from hegemonic centres of power, especially states.

GEG involves a range of actors and processes within and beyond the sovereign state. It involves regulatory processes, mechanisms and organi- sations through which political actors influence environmental outcomes. GEG develops new institutions and new arrangements, such as multi- lateral treaties and conventions, new and more effective international organisations, and new forms of financial mechanisms to account for the dependence of current international regimes on the goodwill of national governments (Lemos and Agrawal, 2006).

GEG includes in addition to the state, actors such as communities, busi- nesses and NGOs. Keys to different forms of environmental governance are the political economic relationships that institutions embody and how these relationships shape identities, actions and outcomes. International accords, national policies and legislation, local decision-making structures, transnational institutions and environmental NGOs are all examples of the forms through which environmental governance can be achieved. These hybrid modes of governance are now becoming common. These approaches have significant potential to help natural systems to recover from environmental degradation (Lemos and Agrawal, 2006).

GEG in minimising CC is challenging because it is distributed over wider groups of stakeholder, none of which can unilaterally influence a deci- sion. CC involves coordination and reconciling multiple actors and interests (Auer, 2000; Lemos and Agrawal, 2006). Important questions such as who should be involved, how agreements should be negotiated and what the rules of engagement are remain intensely divisive and controversial. It is argued by some that because of the rampant inequalities in many rural societies and developing countries, global governance can end up as gover- nance of the many weak by a coterie of powerful elites (Betsill and Bulkeley, 2004).

3. Potential climate-change impacts in Asia

CC is an important political and strategic concern as it can undermine the rapid economic growth rates that have been experienced in Asia over the last few decades. Some of these impacts are discussed below.

3.1 Enhanced natural disasters

CC will increase the intensity and frequency of precipitation, extreme rainfall, droughts, river and inland flooding, storms surges, and environmental health and food security issues in most developing countries in Asia. During 1968–2004, Malaysia experienced 19 natural disasters, which resulted in 1,460 fatalities. The winter monsoon of 2006/2007 and 2007/2008 brought in heavy rainfall and caused severe floods. The Johor flood in 2006 displaced 110,000 people (Herath, 2012).

In China, both the frequency and the intensity of tropical storm surges have increased since the 1960s and it is predicted that there could be more frequent exceptional floods in eastern China in the future (Lewis, 2009). The South China region is susceptible to sea-level rises, which threatens the continued rapid economic growth of coastal areas. Currently the Yellow River delta, the Yangtze River delta and the Pearl River delta are the most vulnerable coastal regions in China.

Floods are frequent in Thailand, and the lower Chao Phraya river basin has suffered from repeated flooding. The 2011 flood affected 77 provinces and a third of the country was under water. In the Philippines the frequency of floods has increased, and in 2008 some 42 floods were reported. The annual flood damage amounted to US$1,829 billion (Herath, 2012).

In India, CC is predicted to increase the severity of flooding, especially in the Godavari and Mahanadi regions on the east coast. Floods are also expected to increase in northwestern India. Extreme precipitation is expected in Central India and Gujarat has experienced severe floods for three consecutive years since 2004, causing large economic losses in the cities (Revi, 2008).

3.2 Impact on food production

India's agricultural sector currently contributes 18 per cent to the country's gross domestic product (GDP) but it provides livelihoods to 60 per cent of the population as well as biomass and ecosystem services (Revi, 2008).

There is an intrinsic link between food security and global changes in climate (Beddington, 2010). CC can damage agricultural systems, reducing yields of crops such as rice, wheat and maize (Lewis, 2009). It can catalyse an agrarian crisis in Asia (Revi, 2008). Malaysia's temperature will rise by 1.5 °C and agriculture will be seriously affected (de Sivar *et al.*, 2009). In Malaysia, since the late 1980s, increased food production and the depletion of natural resources have threatened food security, which can be further exacerbated by CC (de Sivar *et al.*, 2009). Malaysia has nearly 5.5 per cent of arable land under rice, fruits, vegetables and so on, and 17.5 per cent under the permanent crops of oil palm and rubber. About 9.0 per cent of the land in Malaysia is flood prone and the average damage due to floods has been estimated at RM100 million a year (Al Amin *et al.*, 2011).

The Asian Development Bank (2009) estimates that the Philippines, Indonesia, Vietnam and Thailand will experience a drop in rice yield of about 50 per cent by 2100 compared with the 1990 level on average. In Indonesia the decline will be about 34 per cent (Asian Development Bank, 2009). Indonesia receives 15 per cent of its GDP from agriculture and 42 per cent of the labour force are dependent upon the agricultural sector (Arifin, 2012). In Thailand, exports grew by only 0.35 per cent in October 2011 which is the lowest for two years (Khunrattanasiri, 2012).

In India a 10 per cent increase in monsoon precipitation in many regions and a simultaneous precipitation decline of 5–25 per cent in drought-prone central India and a sharp decline in winter rainfall in northern India are projected (Revi, 2008). This will reduce the output of winter wheat and mustard crops in the northwest. In India, food grain production increased four-fold from 50.8 million tons in 1950–1951 to 212 million tons in 2005, but per capita net availability of food grains decreased to 442.8 g/day in 2007 from 494.1 g/day in 2002. At present, 230 million people in rural areas are undernourished; 40 per cent of children under three years of age are underweight and 45 per cent are stunted in growth. By 2020, India's total food requirement will be around 253.3 million tons but the population increase and CC may lead to widespread hunger and starvation (Gahukar, 2009). The hopes for a resurgence in rural agriculture in India are likely to be seriously affected (Revi, 2008).

In Sri Lanka, nearly 77 per cent of paddy production is grown during the wet season. A simulation study showed that a market-led high population growth scenario would decrease rainfall by 17 per cent and increase the irrigation requirement by 23 per cent (de Silva *et al.*, 2007). The same study showed that moderate population growth and a greater concern for the environment would reduce rainfall by 9 per cent and the average irrigation requirement would increase by only 9 per cent (de Silva *et al.*, 2007). Thus rice production will decrease if adequate amelioration of the effects of CC does not occur.

3.3   Climate change and health effects

Severe stress-induced flooding can lead to waterborne diseases, such as malaria and dengue-type epidemics and thus a rise in health expenditure. Stagnant water after floods can increase the incidence of infectious diseases, such as cholera, dysentery, typhoid and paratyphoid (Kolsky, 1999). In the 2006 floods in Malaysia, 20,258 dengue cases with 49 deaths were reported. In 2007, 30,285 dengue cases and 65 deaths were reported (Herath, 2012).

These have occurred mostly in the states of Selangor, Kelantan, Johor and Kuala Lumpur. Serious floods can also stretch the capacity of the health system to cope with the additional workload. Malaria is expected to increase in eastern and northeastern India to western and southern India. Indian cities are major reservoirs of vector-borne diseases, such as malaria and dengue, and the morbidity risk would increase significantly (Revi, 2008). In China, malaria is still an important vector-borne disease, especially in parts of the south. In the Philippines, floods expose people to respiratory infections and skin allergies. Following the 2009 floods in the Philippines, there has been a significant increase in the incidence of leptospirosis due to poor people wading barefoot in flood waters (Juban *et al.*, 2012).

3.4 Differential impacts of climate change on the poor

CC can degrade the resilience of the poor. A study by the United Nations Development Programme (UNDP) (2004) has shown that disaster risks are considerably higher in developing countries because in these countries the poor households lack coping strategies and hence become more vulnerable to environmental risks. However, policy-makers have been ambivalent in conceptualising the centrality of the poor in mitigation and adaptation. Breaking this impasse is a necessary conceptual leap for the policy-maker.

Analysis of data on disasters in 73 nations from 1980 to 2002 shows that richer nations suffer fewer deaths from disasters (Khan, 2006). Countries with robust institutions also suffer fewer fatalities from natural disasters and gastro-intestinal illnesses, with children being most at risk (Zoleta-Nantes, 2002). According to the UNDP, 11 per cent of those exposed to drought, floods and so on are from the developing countries and 53 per cent of them lose their lives (Schipper and Pelling, 2006). Poorer countries also lose more in economic terms. During the 1998 floods in Bangladesh, the government had to borrow US$309 million to finance public spending.

Whether developing countries should also be obliged to reduce their emissions remains controversial due to equity issues. Hence developed countries in the north should either take action to reduce their own emissions or provide the financing necessary to achieve this in developing countries.

4. Environmental governance initiatives to mitigate and adapt to climate-change impacts

There is international pressure to curb GHG emissions but no consensus has yet emerged regarding the best mechanism to cut back on them globally. Policy should focus on mitigation and adaptation to avoid major breakdowns of ecosystems. This is consistent with the burgeoning policy agendas on mitigating and adapting to CC. Adaptation to CC requires a better understanding of the interlinkages between CC and the socioeconomic vulnerabilities of the population. In addition to GEG, regionally/nationally nuanced strategies are required to respond to climate crisis drawing from local experience.

4.1   National policies for adaptation and mitigation of climate change

Multiple subregionally nuanced strategies will be required to respond to climate crisis, drawing on considerable climate experience of coping with uncertainty. All countries must promote awareness, capacity-building, and encouraging organisations and the public to take affirmative action to evaluate and respond to CC. GEG mechanisms can guide national government agencies and the community with regard to how to address CC issues. However, national initiatives are important, especially for CC adaptation.

Many governments in Asia have improved public services and infrastructure, including Integrated River Basin Management, promoting Integrated Flood Management. Malaysia has developed a downscaled hydroclimate model for Sabah and Sarawak. In Johor, multitemporal Radasat–1 images, Landsat ETM+ images, topographical maps and land-use maps provide information about the flood extent and flood depth maps, and both maps show the extent of damage to the environment, lives and property (Huey *et al.*, 2010).

In India the integrated management of the coastal zone, balancing environmental and biodiversity conservation, livelihood and economic development and risk mitigation, are problematic. A series of integrated coastal zone management plans are now in progress. These are important evidence-based mitigation and adaptation measures for coastal India and its cities (Revi, 2008). India instigated post-disaster construction and mitigation programmes after the Orissa supercyclone (1999) and the Kachchh earthquake (2001) and Indian Ocean tsunamis.

By 2025, 70 Indian cities are expected to have a population of more than a million. This will have an important bearing on global climate vulnerability and the potential for mitigation and adaptation. India and China have invested heavily in infrastructure, energy, water and telecommunications for growth. There is a need to re-examine these development trajectories for the mitigation of GHG emissions (Revi, 2008). Most Asian countries were preoccupied with poverty alleviation and social development, and they paid less attention to development of natural hazard risk and mitigation capacity.

Human resources management is a bottleneck in China and Indonesia in disaster management. Juban *et al.* (2012) argue that strengthening human resources is important in the Philippines to avoid the spread of disease. The Malaysian government recognises that institutional capacity for implementation of disaster management is limited.

China, Sri Lanka and Malaysia have adopted measures to ensure food availability in case of disasters. Here minimum food-security quotas are kept that can prepare a nation for natural and social disasters. In Asia, many adaptations have been made by rural communities in response to flooding. Chan and Parker (1996) refer to raising houses on stilts, constructing homes on plinths, livelihood diversification and the mobilisation of community-based support networks to provide shelter and food as some adaptations (Abdelhak *et al.*, 2012).

4.2 Global environmental governance programmes

This section evaluates three GEG mechanisms to control CC – namely, the Kyoto Protocol (KP), the Asia-Pacific Partnership (APP) and Forest certification.

4.2.1 The Kyoto Protocol

The KP is the most comprehensive global environmental governance to reduce GHGs emissions. It includes the 42 members of the Alliance of Small Island States vulnerable to CC, members of the Organisation for Petroleum Exporting Countries and many other developed countries (Yamin, 1998). KP is a complex arrangement involving a range of policy options and varied engagement by multiple levels of governance systems. It is the most disparate and inchoate group of countries ever assembled to tackle a common environmental problem.

The KP mandated the reduction of emissions by developed countries by 5 per cent by 2012. According to the KP, the 37 most industrialised countries of the 146 nations will reduce their GHG emissions below 1990 levels during the first commitment period of 2008 through 2012. This 5 per cent reduction is very small yet it may mean that some countries have to cut their business by as much as 30 per cent to achieve the designated reductions (Rabe, 2007).

The KP also embodies emissions trading, joint implementation and carbon offsetting mechanisms. Through its flexible mechanisms, joint implementation and the Clean Development Mechanism, it has created an arena for new governance experiments, providing ample opportunities for non-state actors. More than a decade after the formation of the United Nations Framework Convention on Climate Change (UNFCCC) the KP was ratified by a sufficient number of states to come into effect in February 2005 (Christoff, 2008).

There is now a consensus that the KP has failed to achieve targeted emissions reductions to mitigate CC. The United States and Australia have declined to ratify. India and China are unwilling to accept binding reduction targets (Kellow, 2010) and have no obligation to reduce emissions. Any long-term reduction of emissions requires the agreement of these countries, including China and India. Developing countries, such as China, which are dependent on coal, may be required to switch to cleaner yet more expensive fuels to limit emissions. This shift may be difficult and the distributional impacts of such policies are an important determinant in their acceptance. These conflicting scenarios have thwarted progress of the KP towards achieving its objectives, thus bringing the process to a standstill.

Another argument is that KP was densely populated with non-state actors with a multitude of interests, capacities and varying contributions to GHG emissions (Andonova, Bersill and Bulkeley, 2009). There was disagreement among these members on the details. There was continuing pressure on the US senate not to ratify the KP. The developed nations of North America and Europe are the most significant emitters of GHGs. However, agreements on the quantum of emissions reduction by different countries were a contentious issue. The success of the KP depends on developing countries coming to agreement on this issue (Rabe, 2007).

The KP focused more on international diplomacy, a process which is inordinately long but seeks the single best system to reduce emissions. An extraordinary effort to determine 'best possible practice' based on political will, institutional capacity and technological transformation vitiated the KP from the more important issues of governance of complex agreements (Hay and Mimura, 2006; Rabe, 2007).

Implementation and consensus received less attention, thwarting progress towards a more flexible and governable mechanism which affected the Asian developing countries in particular. Another issue is that many developing countries lack the technical and institutional capacity to develop and implement policies to complement the KP in minimising GHG emissions.

This is disheartening in the context of the need to reduce emissions by at least 60 per cent over the next 50 years because it is the most significant initiative to reduce CC. Kellow (2006, 2010) argues that having a smaller number of nations (six in the APP) accounting for half of the current emissions provides the opportunity for better policy to be developed, which can overcome some of the problems experienced in previous protocols. The APP is such an initiative.

4.2.2   The Asia-Pacific partnership

The creation of the APP on Clean Development and Climate is another example of GEG (Asselt, 2007). The APP was founded in 2005 (Fujiwara, 2007). It emerged due to the United States' rejection of the KP and the search for alternatives to mitigate CC (Karlsson-Vinkhuyzen and Asselt, 2009). The APP has seven partner countries – Australia, Canada, China, India, Japan, the Republic of Korea and the United States. The initial six partner countries, other than Canada, account for 50 per cent of GHG emissions and 48 per cent of global energy use, and they produce about 65 per cent of the world's coal, 48 per cent of the world's steel, 37 per cent of the world's aluminium and 61 per cent of the world's cement.

It was not a rigid framework for target-setting or trading in GHG emissions but an initiative to coordinate policies and projects for removing barriers to the mitigation of CC. This kind of partnership is more appropriate to the specific resource endowments of the different partners and their unique interests. China considers the APP to promote international cooperation suited to each country (Zhang, 2006). The United States also stresses the national context within the APP to identify technologies and financial arrangements (Connaughton, 2006). The APP has a decentralised structure and flexibility to accommodate divergence in national and sectoral considerations (Fujiwara, 2007). But researchers argue that the AFP is only partially about mitigating CC (Asselt, 2007).

In the APP, participation is voluntary to advance climate objectives but also recognise the need for poverty alleviation, which is most appropriate for developing countries in Asia. This enhanced cooperation meets the increased energy needs and minimises GHG intensities. The APP focuses on cleaner and more efficient technologies (Fujiwara, 2007). It is an initiative that combines sectoral cooperation across countries on the development and deployment of technologies, with sectoral reforms in selected countries for removing barriers to achieving full potential to reduce GHG emissions and energy-efficiency improvements. The former will benefit from the involvement of business whereas the latter will suit government-to-government actions (Connaughton, 2006).

The APP has members who are highly heterogeneous and there are no common legally binding rules or targets coordinated through existing international institutions. The APP is based on a highly decentralised structure that largely relies on the coordination of policies and projects along the lines of the UNFCCC. It depends on contributions from partners to individual projects (Asselt, 2007).

The APP on Clean Development and Climate was an innovation to mitigate CC that can support the KP. The problem is that although the APP is designed to be a public–private partnership, governments appear to dominate the process. It is argued that the APP can obstruct the effectiveness of the KP, and the expected complementarity of the APP with the KP is superficial. But the APP can be a competing alternative that can affect the future shape of the KP (Mcgee and Taplin, 2006). The APP is exerting some influence on UN discussions through its promotion of a sectoral approach. There are fears that moving towards a fragmented legal architecture as in the APP and moving away from binding international agreement can reduce the role of the UN to a simple coordinating mechanism (Asselt, 2007).

4.2.3 Tropical deforestation and forest certification

Deforestation in Southeast Asia is a major factor causing CC. Oil palm in Malaysia has caused forest loss of about 5 million ha (20 per cent reduction of forest land; Fitzherbert *et al.*, 2008; Turner *et al.*, 2008; Wicke *et al.*, 2011). Compared with the estimated extent of primary forests 8,000 years ago, before large-scale human disturbance, relatively little remains intact in Malaysia (11.6 per cent) and Indonesia (25.6 per cent). In 2008, Malaysia's Federal Land Development Authority announced plans to establish oil-palm plantations in Kalimantan (20,000 ha), Aceh (45,000 ha), Papua New Guinea (105,000 ha) and Brazil (100,000 ha). In June 2009, Malaysian oil-palm developers announced plans to establish a 100,000 ha oil-palm planta- tion and an extraction facility in Mindanao in the Philippines. This large expansion of oil palm in forested areas will continue to replace tropical forests.

Over the periods of 1990–2000 and 2000–2005, deforestation rates in Indonesia climbed from 2.3 per cent to 2.7 per cent per year for its pri- mary (undisturbed) forests and from 1.2 per cent to 1.3 per cent per year for its secondary (naturally regenerated) forests. As a result, sustainable forest- management alternatives have emerged to reduce deforestation and ensure that biodiversity and ecosystem values are protected (Stringer, 2006). Many partnerships between multiple forest stakeholders have been developed under the GEG umbrella to reduce deforestation.

Forest certification has been the most important mechanism developed for industrial logging (Pattberg, 2010). Private certification systems are the most visible GEG forest-management system. Forest certification provides a means by which producers who meet sustainable forest standards can identify their products in the marketplace, receiving greater market access and higher prices for their products (Gullison, 2003). Forest certification has grown significantly in the last two decades. By 2002, there were more than 50 forest-certification systems around the world. Forest certification has improved governance because it is based on market principles.

The Forest Stewardship Council (FSC) is the only global certification with wide geographical coverage. FSC-accredited certifying organisations have more forests in tropical countries where biodiversity conservation needs are greatest. FSC standards have the greatest support from the environmental and social NGOs (Gullison, 2003). FSC has made the greatest inroads in temperate developed countries, certifying nearly seven times as much for- est in Europe and North America. The achievements in Asia have not been as significant.

Major problems have been identified in forest certification. The FSC has little support in the private sector, especially in Asia, because it does not recognise the political and legal difficulties in operating in the Asian region. If the FSC is to have a positive impact in tropical countries such as Malaysia and Indonesia, which are megabiodiversity countries, certification must ensure more tangible benefits and simultaneously reduce the costs of certification.

Mechanisms that mitigate GHG emissions via forest conservation have been considered to be a cost-effective approach to protect biodiversity (Ghazoul, 2001). However, the costs of conservation are spatially heterogeneous across the globe. The lowland rainforests of Southeast Asia represent a unique nexus of large carbon stores, imperilled biodiversity, large stores of timber, and high potential for conversion to oil-palm plantations, making this region one where understanding the costs of conservation is critical (Fisher *et al.*, 2011). Previous studies have underestimated the gap between conservation costs and conversion benefits in Southeast Asia. Using detailed logging records, cost data and species-specific timber auction prices from Borneo, it has been shown that the profitability of logging or the potential profits from subsequent conversion to palm-oil production exceeds revenues from a global carbon market and other ecosystem-service payment mechanisms. Thus the conservation community faces a difficult task in funding to protect the remaining lowland primary forests in Southeast Asia.

Another major weakness of forest certification is that they have a product-orientation focus. Kloosner (2005) found that the retailer focus has limited the spread of forest certification among medium-sized, small and community forest management operations. This will affect equity in the distribution of benefits from forest-certification schemes. This is very important for Asian countries that are mostly producers rather than purchasers. Forest certification has failed in the watchdog function and raising awareness among the general population in the Asian countries, which is critically important.

Scientists have their own tales of woe regarding certification processes. They claim that there is inadequate attention given to biodiversity under forest-certification programmes (Bennett, 2000). They argue that criteria are too general and need to be expanded to accommodate the effects of logging on forest biodiversity (Bennett, 2000). The greatest attention should have been on the magabiodiversity countries, such as Malaysia, Indonesia and Papua New Guinea. Further, ecological sustainability is beyond the capacity of managers of many developing countries in Asia. Many tropical forest managers are constrained in their actions by inadequate technical, financial and human resources, which affect their involvement in forest certification.

In developing countries, lobby groups and powerful interest groups have significant influence on the political process, which can undermine forest certification. Bribery and corruption in the logging industry are rife in Indonesia. Forest-certification schemes provide multilateral agreement on limiting deforestation, and rules to balance the social, economic and ecological values of forest resources. But these programmes in tropical countries have poor governance and high levels of corruption. The FSC's global forestry-certification system may have some anti-corruption effects in countries where corruption is not systemic. In many Southeast Asian countries, however, corruption is systemic. Indonesia is a good example. Corruption in the forest sector of Swat, Pakistan, is another example (Pellegrini, 2007). Forest certification is not feasible if the overall institutional environment is weak. Mitigating corruption in corruption-ridden forest-management systems might enhance the chances of success of forest certification but currently corruption remains a major stumbling block for progress.

5. Conclusion

There is a great degree of awareness of the need for the integration of CC and disaster risk because they are the two sides of the same coin. Global initiatives still appear to vitiate policy-makers because the need for all parties to work collectively has not been forthcoming. Unilateral decisions except in the case of local adaptation and mitigation cannot make any dent in the global problem. The GEG initiatives, such as the KP, were divisive and controversial due to the primacy of national objectives. The United States and Australia did not ratify the KP. How to reconcile national and international priorities remains problematic and challenging. Another issue is that CC is guided by global cooperation but disasters related to risk are taken at the national and subnational levels. National disaster risk-management institutions can undertake adaptation measures for CC. Recent experience also shows that the reduction of GHG will not automatically follow international accords. More serious initiatives post Kyoto must be initiated with greater global cooperation if CC is to be addressed successfully.

Abbreviations

APP	Asia Pacific Partnership
CC	climate change
FSC	Forest Stewardship Council
GDP	gross domestic product
GEG	global environmental governance
GHG	greenhouse gases
IPCC	Intergovernmental Panel on Climate Change
KP	Kyoto Protocol
NGO	non-governmental organisation
UN	United Nations
UNDP	United Nations Development Programme
UNFCCC	United Nations Framework Convention on Climate Change

References

S. Abdelhak, J. Sulaiman, J. and S. Mohd (2012) 'Poverty among Rural Communities in Kelantan and Terengganu: The Role of Institutions, Farmers' Risk Management and Coping Strategies', *Journal of Applied Sciences*, 12, 125–135.

A.Q. Al-Amin, W. Leal, J.M. Trinxeria, A.H. Jaafar and Z.A. Ghani (2011) 'Assessing the Impacts of Climate Change in the Malaysian Agriculture Sector and Its Influences in Investment Decision', *Middle-East Journal of Scientific Research*, 7, 225–234.

L.B. Andonova, M.M. Bersill, H. Bulkeley (2009) 'Transnational Climate Governance', *Global Environmental Politics*, 9, 52–73.

B. Arifin (2012) 'Increasing Environmental Risks and Food Security in Indonesia, In Impact of Increasing Flood Risk and Health Security in South East Asia', Proceedings of the International Symposium on impacts of increasing Flood Risk on food and Health Security in southeast Asia, March 1, 2012, Kyoto, Japan, pp. 71–84.

Asian Development Bank (2009) *The Economics of Climate Change in Southeast Asia: A Regional Review* (Manila: ADB).

H.V. Asselt (2007) 'From UN-ity to Diversity? The UNFCCC, the Asia-Pacific Partnership and the Future of International Law on Climate Change', *Carbon and Climate Law Review*, 1, 17–28.

M.R. Auer (2000) 'Who Participates in Global Environment Governance? partial Answers from International Relations Theory', *Policy Sciences*, 33, 155–180.

J. Beddington (2010) 'Food Security: Contributions from Science to a New and Greener Revolution', *Philosophical Transactions of the Royal Society*, 365, 61–71.

N. Beg, J.C. Morlot, O. Davidson, Y. Afrane-Okesse, L.A. Tyani, A.F. Denton, Y. Sokona, J.P. Thomas, E. Lèbre La Rovere, J.K. Parikh and A.A. Rahman (2002) 'Linkages between Climate Change and Sustainable Development', *Climate Policy*, 2, 129–144.

E.L. Bennett (2000) 'Timber certification: Where Is the Voice of the Biologist?' *Conservation Biology*, 14, 921–923.

M.M. Betsill and H. Bulkeley (2004) 'Transnational Networks and Global Environment Governance: The Cities for Climate Change Program', *International Studies Quarterly*, 48, 471–493.

F. Biermann and P. Pattberg (2008) 'Global Environment Governance taking Stock, Moving Forward', *Annual Review of Environmental Resources*, 33, 277–294.

N.W. Chan and D.J. Parker (1996) 'Response to Dynamic Flood Hazard Factors in Peninsular Malaysia', *The Geographical Journal*, 162, 313–325.

P. Christoff (2008) 'The Bali Map: Climate Change, COP 13 and beyond', *Environmental Politics*, 17, 466–472.

Commission on Global Environmental Governance (1995) *The Report of the Commission on Global Environmental Governance* (Oxford: Commission on Global Environmental Governance).

J.L. Connaughton (2006) Testimony of James L. Connaughton, Chairman, Council of Environmental Quality, Executive Office of the President, before the United Sates Senate, Committee on Environment and Public Works, September 20.

C.S. de Silva, E.K. Wetherhead, J.W. Knox and J.A. Rodriguez-Diaz (2007) 'Predicting the Impacts of Climate Change: A Case Study of Paddy Irrigation Water Requirements in Sri Lanka', *Agro Cultural Water Management*, 1–2, 19–29.

B. Fisher, D P. Edwards, X. Giam and D. Wilcove (2011) 'The High Costs of Conserving Southeast Asia's Rainforest', *Frontiers in Ecology and the Environment*, 9, 329–334.

E.B. Fitzherbert, M.J. Struebig and A. Morel (2008) 'How Will Palm Oil Expansion Affect Biodiversity', *Trends in Ecology and Evolution*, 23, 538–545.

N. Fujiwara (2007) *The Asia Pacific Partnership on Clean Development and Climate: What It Is and What It Is Not*, Policy Brief No. 144, (Brussels: Centre for European Policy Studies).

J. Ghazoul (2001) 'Barriers to Biodiversity Conservation in Forest Certification', *Conservation Biology*, 15, 315–317.

R.T. Guhakar, (2009) 'Sustainable Agriculture in India: Current Situation and Future Needs, *International Journal of Agricultural Sciences*, 5, 1–7.

R.E. Gullison (2003) 'Does Forest Certification Conserve Biodiversity?', *Oryx*, 37, 153–165.

J. Hay, and N. Mimura (2006) 'Supporting Climate Change Vulnerability and Adaptation Assessments in the Asia-Pacific Region: An Example of Sustainability Science', *Sustainability Science*, 1, 23–35.

G. Herath (2012) 'Sustainable Agricultural Development in Malaysia with Special Reference to Food and Health Security and National Disasters', In Impact of Increasing Flood Risk and Health Security in South East Asia, Proceedings of the International Symposium on Impacts of Increasing Flood Risk on Food and Health Security in Southeast Asia, March 1, 2012, Kyoto, Japan, 85–92.

T.T. Huey, A.L. Ibrahim, M.S. Saayonand and M.Z.A. Rahman (2010) 'Remote Sensing Methods for Mapping Flood-Prone Areas', Paper Presented at the Asian Association of Remote Sensing, November 1–5, 2010, Vietnam.

IPCC (1995) *The Science of Climate Change: Summary for Policy Makers*, (WMO Geneva), 44–45.

N. Juban (2012) 'The Epidemiology of Disasters: health Effects of Disasters in the Philippines', In Impact of Increasing Flood Risk and Health Security in South East Asia, Proceedings of the International Symposium on Impacts of Increasing Flood Risk on Food and Health Security in Southeast Asia, March 1, 2012, Kyoto, Japan, 53–60.

S.I. Karlsson-Vinkhuyze and H.V. Asselt (2009) 'Introduction: Exploring and Explaining the Asia-Pacific partnership on clean development and climate, international environmental agreements', *Politics, Law and Economics*, 9, 195–211.

A. Kellow (2006) 'A Process of Negotiating Multilateral Environmental Agreements? The Asia Pacific climate partnership beyond Kyoto', *Australian Journal of International Affairs*, 60, 287–303.

A. Kellow (2010) 'Is the Asia-Pacific partnership a Viable Alternative to Kyoto, WIRES Climate', *Climate Change*, 1, 10–15.

M.E. Khan (2006) 'The Death Toll from Natural Disasters: The Role of Income, Geography and Institutions', *The Review of Economics and Statistics*, 87, 271–284.

W. Khunrattanasiri (2012) 'Increasing Flood Risk in Thailand: Causes an Solutions', In Impact of Increasing Flood Risk and Health Security in South East Asia, Proceedings of the International Symposium on Impacts of Increasing Flood Risk on Food and Health Security in Southeast Asia, March 1, 2012, Kyoto, Japan, 29–38.

D. Kloosner (2005) 'Environmental Certification of Forests: The Evolution of environmental Governance in a Commodity Network', *Journal of Rural Studies*, 21, 403–417.

P. Kolsky (1999) 'Engineers and Urban Malaria: Part of The Solution or Part of the Problem?', *Environment and Urbanisation*, 1, 159–164.

J.I. Lewis (2009) 'Climate Change and Security: Examining China's Challenges in a Warming World', *International Affairs*, 85, 1195–1213.

M.C. Lemos and A. Agrawal (2006) 'Environmental governance', *Annual Review of Environmental Resources*, 31, 297–325.

J. Mcgee and R. Taplin (2006) 'The Asia Pacific Partnership on Clean Development and Climate: A Complement or Competitor to the Kyoto Protocol?', *Global Change, Peace and Security*, 18, 173–192.

L. Schipper and M.Pelliong (2006) 'Disaster Risk, Climate Change and International Development: Scope for, and Challenges to, Integration', *Disasters*, 30, 19–38.

A.B. Shrestha, C.P. Wake, P.A. Mayewsi and J.E. Dibb (1999) 'Maximum Temperature Trends In The Himalayan And Its Vicinity: An Analysis Based on the Temperature Record from Nepal for the Period 1971–1994', *Journal of Climate*, 12, 2775–2789.

P. Pattberg (2010) 'Public-private Partnerships in Global Climate Governance', *Wiley Interdisciplinary Reviews: Climate Change*, 1, 279–287.

I. Pellegrini (2007) 'The Rule of the Jungle in Pakistan: A Case Study On Corruption and Forest Management in Swat', FEEM Working Paper No. 91 (Milan: Fondazione Eni Enrico Mattei).

B. Rabe (2007) 'Beyond Kyoto: Climate Change Policy in Multilevel Governance Systems', *Governance*, 20, 424–444.

H.A. Raman (2009) 'Global Climate Change and Its Effect on Human Habitat and Environment in Malaysia', *Malaysian Journal of Environmental Management*, 10, 17–32.

A. Revi (2008) 'Climate Change Risk: An Adaptation and Mitigation Agenda for Indian Cities', *Environment and Urbanisation*, 20, 207–229.

B. Shui, and R.C. Harriss (2006) 'The Role of CO_2 Embodiment in US-China Trade', *Energy Policy*, 34, 4063–4069.

M.C. Smouts (1998) 'The Proper Use of Governance in International Relations', *International Social Science Journal*, 50, 81–89.

C. Stringer (2006) 'Forest Certification and Changing Global Commodity Chains', *Journal of Economic Geography*, 6, 701–722.

E.C. Turner, J.L. Snaddon, T.M. Fayle and W.A. Foster (2008) 'Oil Palm Research in Context: identifying the Need for Biodiversity Assessment, http://www.plosone.org, date accessed 10 March 2013.

UNDP (2004) *Reducing Disaster Risk: A Challenge for Development* (New York: UNDP).

B. Wicke, R. Sikkema, V. Dornburg and A. Faaij (2011) 'Exploring Land Use Changes and the Role of Palm Oil Production in Indonesia and Malaysia'. *Land Use Policy*, 28, 193–206.

F. Yamin (1998) 'The Kyoto Protocol: Origins, Assessment and Future Challenges', Review of European Community & International Environmental Law, 7(2), 113–127.

H. Zhang (2006) 'China's Position on International Climate Change Negotiations: Continuity and Change', *Journal of International Peace*, 3, 1–20.

D.B. Zoleta-Nantes (2002) 'Differential Impacts of Flood Hazard Risk among Street Children, the Urban Poor and Residents of Wealthy Neighbourhoods in Metro Manila, Philippines', *Mitigation and Adaptation Strategies for Global Change*, 7, 239–266.

Part II

Preconditions of Good Governance and the Role of Different Sectors

6
Approaches to Climate Change Adaptation: A Case Study of Agricultural Initiatives in Japan

Izumi Tsurita, S.V.R.K Prabhakar and Daisuke Sano

1. Introduction

Climate change (CC) impacts are threatening around the globe and they are not limited to developing countries. CC adaptation in developing countries is mostly focused on community-based adaptation to help the poorest and the most vulnerable societies by empowering communities to become resilient to CC impacts (Huq and Reid, 2007; Ayers and Huq, 2009). In contrast, developed countries are presumed to have a low vulnerability to CC (Ford and Berrang-Ford, 2011). As a result, not many studies are conducted in the developed countries to see the effectiveness of policies and actions taken for CC adaptation. However, as the adaptation is location specific and solutions and approaches are diverse (Agrawal, 2010), it is worthwhile understanding cases in developed countries.

Impacts on rice production are anticipated in Japan and several approaches are adopted to maintain the quality and quantity of rice production. Study of these approaches could yield useful lessons for other rice-producing countries in the Asia-Pacific region. Keeping this in mind, this chapter examines the adaptation measures taken in the agricultural sector in Japan, focusing on rice production, and it identifies critical messages and gaps with a view to finding effective ways to meeting Japan's adaptation needs.

1.1 Observed climate change impacts in Japan

Due to increasing temperatures and less or too much rain, symptoms, such as chalky grain, have been observed in Japan (Committee on Impacts of Global Warming and Adaptation, 2008). Chalky grain is the murky colour of rice grain which is considered to be poor quality. First-class quality rice is the determinant of the highest rice price, and its criteria are mostly based on the appearance of harvested rice, such as size and colour, as well as the amount of grain moisture. Negative impacts on the quality of crops have a direct effect on the market price, sales and the income of the farmers. In addition to the direct impact of changing temperatures on crops, shifts in crop pests have also been observed. For example, grain damage caused by the infestation of rice leaf bug, rice bug and mirid bug has become more frequent (National Agricultural Research Centre for Tohoku Region, 2006). An increase in the infestation period of rice stem borer (*Chilo suppressalis*) was observed with the extended duration of warm temperatures during autumn (Seino, 2008).

Negative climatic impacts on rice have increased since 2000, especially in the northwest of Japan, including the Niigata prefecture, and in the southern part of Japan, including the Miyazaki prefecture, where the average temperatures during the grain-filling period are increasing (Seino, 2008). Niigata is located on the island of Honshu on a plain between the coast of the sea of Japan and Echigo Mountains. With a humid subtropical climate, Niigata receives about 1,821 mm precipitation per year with rainfall during July to November and snowfall during December to March. The average daily mean temperature (about 13.9 °C), the average high temperature (about 30.6 °C in August) and the average low temperature (0.1 °C in February) in Niigata vary significantly for its four distinctive seasons (Japan Meteorological Agency, 2012a). Miyazaki is located on the island of Kyushu surrounded by the Pacific Ocean. With the warm Japan Current, it enjoys a mild subtropical climate. Miyazaki receives about 2,508.5 mm precipitation per year with heavy rain during June to September. The average daily mean temperature (about 17.4 °C), the average high temperature (about 31.4 °C in July) and the average low temperature (2.6 °C in January) in Miyazaki vary in four distinctive seasons (Japan Meteorological Agency, 2012b). Such seasonal variations in temperature have clear implications for rice production and the quality of rice.

Niigata experienced degraded rice quality in 1999 due to the abnormal summer temperatures and in 2004 due to the large number of typhoons. In 2010, the first-class quality rice in Niigata was about 19.7 per cent which is the lowest ever, affecting farmers' income (Ministry of Agriculture, Forestry and Fisheries (MAFF), Japan, 2010a). The average minimum temperature in Niigata was 25.9 °C during August 2010, which was just 0.1 °C below the threshold temperature for chalky grains (Ouchi, 2010). In 2007, Miyazaki was affected by the *foehn* phenomenon after the typhoon, resulting in an extremely low rate of first-grade rice. The *foehn* phenomenon is known to be an extremely dry and hot local air which occurs when the humid wind flows over the mountain or when the hot air in the upper levels falls along the leeward side of the mountain. Severe impacts on rice quality in Niigata and Miyazaki encouraged these prefectures to respond rapidly to climatic variability.

1.2 Future climate change projections in Japan

In the medium term until the 2050s, a gradual increase in temperature and carbon dioxide (CO_2) are predicted to benefit rice production in Japan (Yokozawa and Iizumi, 2009). Rice yields may increase for a while because of the reduced impact of cold summers and the CO_2 fertilisation effect which is caused by increasing photosynthesis with increasing CO_2 concentrations (Seino, 2008).

In Hokkaido, rice quality and yield are likely to increase if the geographical suitability for rice production and farmers' capacity permits (MAFF, Japan, 2008). However, such an increase in yield may not occur in eastern and southern Japan since the temperature increase in those areas is already rapid, particularly in summer and autumn. An increase in total production will peak once the temperature exceeds the optimum range for growing rice, especially during the flowering and grain-filling periods (Yokozawa and Iizumi, 2009).

In the long term, towards the end of this century, both the quality and the quantity of rice in the whole of Japan are likely to decline due to the compound stresses caused by CC (Committee on Impacts of Global Warming and Adaptation, 2008).

2. Research objectives

This chapter aims to examine how a developed country such as Japan is currently responding to CC by studying various initiatives in the agriculture sector on a case study basis. The objective is to know what is being done for CC adaptation in Japan and what more can be done.

3. Research methods

A literature review, a consultation workshop with experts and an open-ended interview with personnel at prefectural-level government, agricultural research institutes and extension agencies were conducted to examine how Japan is currently responding to CC in the agriculture sector. The reviewed literature includes reports and documents published by the national and local governments and research institutes, peer-reviewed journals, academic remarks and other technical books.

A policy consultation workshop entitled Adaptation in Agriculture and Water Sectors in Japan and Its Relevance for Developing Countries in the Asia Pacific was held on 13 July 2010 in Pacifico Yokohama, Japan. It was a part of the annual International Forum for Sustainable Asia and the Pacific organised by the Institute for Global Environmental Strategies.

Interviews were conducted in two case-study locations: the Niigata and Miyazaki prefectures in 2010. These locations were selected based on the publications and announcements of the MAFF, Japan, and the Ministry of

Foreign Affairs of Japan. The presence of CC adaptation-relevant initiatives in the agriculture sector was the main criterion. Interviews were conducted with the officers of the Agricultural Production Section and Research Management Office of Niigata Prefectural Government, the Niigata Agricultural Research Institute, the Global Warming Research Centre for Agriculture and Fishery of Miyazaki, the Prefecture Agricultural Experiment Station and the Miyazaki Prefecture Central Agricultural Development and Extension Centre.

4. Findings

4.1 Adaptation strategies at the national level

4.1.1 Overall adaptation strategies

Japan's leading adaptation scientists group, the Commission on Direction of Climate Change Adaptation, applies the definition of adaptation from the Intergovernmental Panel on Climate Change (IPCC), which is 'Adjustment in natural or human systems in response to actual or expected climatic stimuli or their effects, which moderates harm or exploits beneficial opportunities' (IPCC, 2001, p. 750).

Our review of existing policy documents in Japan suggested that if one only looks for the term 'adaptation to climate change', such policies in Japan would be extremely limited. Table 6.1 lists basic projects, strategies and plans that are either directly addressing CC adaptation or have adaptation benefits. Ministries such as the Ministry of the Environment, the MAFF, the Ministry of Education, Culture, Sports, Science and Technology (MEXT), the Ministry of Land, Infrastructure and Transport (MLIT), and the Ministry of Health, Labour and Welfare (MHLW) are actively involved in implementing current adaptation policies.

It is clear from Table 6.1 that the Japanese government has put special emphasis on research and technology development especially for predicting future climate and assessing the potential impacts.

These approaches are essential for setting the basis for strategic plans and effective implementation. However, as the Organisation for Economic Co-operation and Development's (OECD's) environmental performance review of Japan pointed out, a comprehensive national legally binding adaptation strategy is yet to be established in Japan (OECD, 2010).

4.1.2 Adaptation in the agriculture sector

Several past developmental initiatives in the agriculture sector with adaptation benefits already exist in Japan, including the development of an irrigation network with huge support from public works (Council of Food, Agriculture and Rural Area Policies, 2006). New crop varieties and farming

Table 6.1   Major adaptation initiatives in various sectors in Japan

Sectors	Initiatives
Cross-cutting	• Climate Change Symposium 2007 (Cabinet Office) • Reinforcement of weather and climate data observation using Greenhouse gases Observing SATellite (Japan Meteorology Agency; JMA) • Comprehensive research for global warming 2005–2009 and report on Wise adaptation 2008 (MOE) • Establishment of data integration and analysis system (MEXT) • Kakushin Project for future forecast of CC, 2007–2011 (MEXT) • Direction of CC adaptation (MOE 2010)
Agriculture and water	• Annual CC impact report by product item (climate change adaptation report by item 2007) (MAFF, Japan, 2007) • Technology development and impact assessment of global warming for agriculture and fishery sectors, 2002–2005 (MAFF) • Water-shortage information portal from 2002 (MLIT, MHLW, MAFF and JMA) • Interim report on integrated water-resource management addressing CC and other risks 2008 (MLIT) • Comprehensive strategy for the CC 2008 (MAFF) • Vision of aqueduct 2004 (MHLW)
Ecosystems	• Project of monitoring site 1,000 from 2003 (MOE) • Comprehension of high-risk mountainous disaster area and research on proactive measurements for mountainous areas • Understanding high-risk mountainous disaster area (Forestry Agency) • Third strategy on biodiversity 2007 • Comprehensive strategy for CC, 2008 (MAFF)
Disaster management	• Research on advanced technology in risk assessment and observation of severe weather, 2009–2013 (MLIT and JMA) • Abnormal climate report, 2005 (JMA) • Review of water disaster risk-assessment method 2008 and how-to information about designing a Hazard Map, 2005 (MLIT) • Disaster Risk Information Platform 2008 (MEXT) • Research on landslide and wind/water disaster occurrence prediction by multiparameter radar, 2007 (MEXT)

Table 6.1   (Continued)

Sectors	Initiatives
	• Plan for CC adaptation in water disaster-management sectors, 2008, recommendation in relation to the landslides in midterm vision, 2007, and Urban and Rural Transport Strategies, 2005 (MLIT) • Strategy for the 21st-century environmental nation, 2007 (MOE)
Health	• Research on global health issues including 'annual research on health impact by climate change from 2000' and 'water management to cope with climate change 2009'(MHLW)

Sources: Developed from Council of Science and Technology Policy (2010) and Pacific Consultants Co LTD (2010).

practices have been developed over the years to cope with the changing natural environment and changing farming and consumer markets. For example, the Hokkaido region was believed to be too cold for growing rice in the 19th century. Rice production was the lowest priority for agricultural development during that time. However, in response to the demand for rice, crop varieties and cultivation techniques were improved significantly after the mid-20th century with the effort of local practitioners. Now the rice yield and rice quality in Hokkaido are ranked as some of the highest in the nation (Hokkaido Agricultural Policy Department, 2009).

In terms of specific responses to CC, the MAFF established a Global Warming Taskforce with a clear emphasis on greenhouse gas (GHG) mitigation in October 2009 (MAFF, Japan, 2010b). The taskforce is headed by the minister of MAFF and managed by the Environment and Biomass Policy Division of the minister's secretariat, the MAFF. The main objectives for the taskforce are: (i) to reduce GHG emissions from agricultural activities, (ii) to absorb GHG emissions from effective forest and land management and (iii) to reduce GHGs emissions by facilitating biomass energy and renewable energies (MAFF, Japan, 2010b).

Adaptation strategies stated in Chapter 3 of 'MAFF Comprehensive Strategy for the Climate Change (decided in June 2007 and revised in July 2008)' indicates the promotion of adaptation through training and sharing present technologies, and developing, demonstrating, introducing and reviewing technologies (MAFF, Japan, 2008). The MAFF has conducted CC impact assessment in collaboration with counterparts in each prefecture to facilitate the technical assistance for agricultural production. Impacts on major crops (symptoms such as chalky rice grain) and possible adaptation measures proposed by the national and prefectural agricultural research institutes were compiled (MAFF, Japan, 2007). The development of heat- and disease-tolerant agricultural crops has been initiated at the national- and prefectural-level research centres. In the case of rice production, *Nikomaru*, the high-temperature-resistant rice variety developed by the National Agriculture and Food Research Organisation, is already being promoted around the Kyusyu area (MAFF, Japan, 2008). However, no direct national regulations were established for CC adaptation by the MAFF and most approaches focused on research and related capacity-building. As a consequence, adaptation policy measures are not so apparent in the form of incentives, regulations or approaches.

4.2 Adaptation strategies at the prefectural and local level

Climate-related issues are handled by multiple divisions in the Niigata and Miyazaki prefectures, mostly by the social and environmental divisions that deal with livelihoods, health and water management and not by the agricultural division. In other prefectures, agricultural divisions/offices are responding to such needs as a priority, depending on the awareness, understanding, resources available, willingness of the prefecture and the severity of the climatic impacts that these prefectures have perceived.

4.2.1 Niigata

The Global Environment Countermeasure Office in Environmental Planning Division of Residents' Life Environment Department is in charge of CC issues. It has developed the Niigata Global Warming Regional Countermeasure Promotion Plan 2009 (*Niigataken chikyu ondanka taisaku chiiki suishin keikaku*, 2009) (Niigata Prefecture, 2009a). This focuses on GHG mitigation, and only one paragraph out of 76 pages mentions adaptation.

The Agriculture, Forestry and Fishery Department issued the 'Agriculture and fishery vision 2006 (*Nourin Suisan Vision*, 2006)' (Niigata Prefecture, 2006) based on the Niigata prefecture's overall basic development plan towards 2020, namely 'Dream developing policy plan 2006 (Yume Okoshi Seisaku Plan, 2006; revised in 2009)' (Niigata Prefecture, 2009b). Only a few components considering CC have been included in the vision – that is, environmental protection through soil, air and water purification, and sustainable forest management.

To respond to the degrading rice quality due to increasing temperatures, the development of a new rice variety, *Cho-Koshihikari*, was started in April 2010 as a project of the Agriculture, Forestry, and Fisheries Department in the Niigata prefecture. (In Japanese, *Cho* means super and *Koshihikari* is one of the rice varieties which is popularly planted in Japan today. The superior flavour of *Koshihikari* is widely accepted by consumers and thus, by adding *Cho*, it is intended to develop super *Koshihikari*.) The prefecture aims to create a heat-tolerant, late-maturing rice variety which is as tasty as that of the existing Niigata-brand mid-maturing rice variety *Koshihikari* or the early-maturing variety, *Koshiibuki*. The project is fully funded by the prefectural government (approximately 60 per cent of the total project costs) and the Japan Agricultural Cooperatives (JA) group, and part of the revenues comes from rice seed sales. A premium was added to rice seeds that farmers buy from the seed companies. Farmers are informed of this in the seed order form. Since there is too much dependence on the existing variety (*Koshihikari*), its development will diversify the rice varieties planted in the region and reduce the damage risk from the fluctuation in temperature during the critical rice-cultivation period.

The development of the early-maturing rice variety, *Koshiibuki*, was also initiated to deal with changing climatic conditions after the prefecture experienced cool weather in 1993, drought and heat in 1994, and hot night temperatures in 1999. About 55 per cent of total rice production failed to be graded as first-class quality in 1999 (as noted by the government officials of the Niigata prefecture during the interview). Even before the wide acknowledgement of CC, research had been conducted to cope with fluctuating temperatures. With its continuous efforts, in 2001, the Niigata prefecture started commercial production of *Koshiibuki*, the rice variety that is tolerant to hot temperatures during the grain-filling period with comparable quality and taste of *Koshihikari*. Usually the prefecture's agricultural research centre develops and tests suitable varieties of crops (including rice), registers them as recommended varieties, and introduces them into commercial-scale production. This process often takes 10–15 years.

The prefectural government also provides guidance to farmers through extension services. It recommends that farmers fix the planting date on 10 May each year (about 7–10 days later than the conventional date) in order to delay the rice spike emergence period. It also recommends that farmers plant *Koshihikari* in 70 per cent of their rice paddy and other varieties in 30 per cent to reduce yield and quality decline from epidemics and widespread crop loss due to monoculture (personal communication with Agricultural Production Section and Research Management Office of Niigata Prefectural Government). The latter recommendation has not been welcomed by farmers because of the high market value of *Koshihikari*, even though they could reduce the risks of dependency on a single variety. Hence the prefecture has to widen its genetic base of high-quality rice varieties so that farmers can cultivate different genotypes while benefiting from the same or similar high-quality rice that the prefecture is known for.

4.2.2   Miyazaki

The Environmental Administration Division of the Environment and the Forest Department are in charge of the prefectural CC policies. 'Miyazaki global warming countermeasure implementation plan (Miyazaki chikyu ondanka taisaku jikko keikaku)' was issued in 2000 (Miyazaki prefecture,

2012). Similar to the Niigata prefecture, most activities promoted by the Environmental Administration Division focuses on mitigation.

Independently of the above plan, the Agricultural and Fishery Administration Department has recently issued three major plans to promote research into global warming, to showcase global warming adaptive agricultural management and to develop mitigation technologies, including bioenergy such as poultry manure power plants (Global Warming Research Centre for Agriculture and Fishery, 2008). In 2008, the prefecture individually established the Global Warming Research Centre for Agriculture and Fishery under the supervision of the Agricultural and Fishery Administration Department. This centre is one of very few research institutes specifically established for the purpose of studying global warming for agriculture at the prefectural level in Japan. By implementing pilot model projects for CC, the research centre is supporting the implementation of the Production Area Structural Reform Plan to Cope with Global Warming in the Miyazaki Prefecture (*Miyazakiken chikyu ondanka taio sanchi kozo kaikaku keikaku*)' for the Agricultural and Fishery Administration Department. Since the Miyazaki prefecture is one of the major citrus- and mango-producing areas, the government officials we interviewed believed that global warming could open up new opportunities for producing alternative crops, such as tropical fruits, and diversifying the agricultural products of the Miyazaki prefecture. A few months after the interview, the Miyazaki prefecture announced that it would test-grow lychee and jujube to revitalise the region's high-quality local products. Warmer temperature may also reduce the fuel consumption involved in greenhouse mango production since approximately 6 l of fuel is currently consumed to produce one mango.

As mentioned before, Miyazaki is located in southern Japan so its climate is warmer than that of many other prefectures and it is experiencing degrading rice quality. Therefore the development of a heat-resistant rice variety has been initiated. The prefecture is growing two major rice varieties: *Koshihikari* (early-maturing variety planted in February, 3,000 ha in the prefecture) and *Hinohikari* (mid-maturing variety, 1,000 ha). The quality of both rice varieties was affected by the changing climate in the form of an increase in chalky grain, pests, viruses and invasive species (personal communication with Central Miyazaki Agricultural Extension Centre). *Koshihikari* is susceptible to rice blast and wind, and its flowering coincides with the high typhoon rainfall. The new early-maturing variety currently being developed (*Miyazaki* No.45) will have a shorter stem and hence resistant to lodging (lodging refers to collapse of rice plants due to unfavourable weather and nutrient management). The development of this variety is supported by the prefecture and is in the field experimental stage, to be ready for adoption in a few years. Funded by the national government, *Hinohikari* was developed by the Agricultural Research Institute in the Miyazaki prefecture for producing good-flavoured rice even in a warm climate, such as in the Kyushu region (MAFF, Japan, 2009).

4.2.3 The role of cooperatives and local extension agencies

In Japan, extension agencies and JA assist farmers at the local level. Extension agents work under the prefectural administration, and this is linked with the agricultural department of the prefecture and the prefectural agricultural research centre. The JA is a Japanese farmers association initiated under the Japan Agricultural Cooperatives Law 1947 by the General Headquarters of the Allied Forces of the United States. The idea of JA was to formulate an autonomous agricultural organisation for farmers in order to ensure equal rights to peasants. However, under the severe food shortages after World War II, it was difficult to make JA completely independent from the government. In the post-war era, the JA has been playing a middle-man role between the government and farmers. Although most of the farmers are the members of JA, the organisation itself is partly managed by the farmers (Yamashita, 2009).

Technicians at the extension centre give advice and training to farmers according to guidance from the research centres. Due to the limited number of staff, the extension agency may not always be able to cover all of the municipalities so they often collaborate with local JA branches to reach farmers. In most instances the extension centre advises JA agents to instruct lead farmers in the district, and lead farmers disseminate the instructions to all other farmers. In Miyazaki, extension agents and JA provide all farmers with information about farming techniques for efficient water and fertilizer management in the critical stages of cultivating rice. In addition, they organise training and workshops after harvesting (off season). In the year of extreme weather, more training and meetings were held for review and improvement. Interviews with extension agents in the Central Miyazaki Agricultural Extension Centre revealed that the extension agents do not have a specific mandate to initiate work on CC. Most of their efforts are dedicated to reaching out to farmers in the area to deal with regular crop-production-related issues, such as pest control and irrigation, and no special training CC adaptation has been imparted to the extension agents. Similarly, JA does not have a specific division in charge of CC matters. Our interviews also did not reveal the existence of specific training programmes for farmers to deal with the vagaries of the climate.

5. Analysis and discussions

Figure 6.1 shows the analysis of adaptation actions at each level of government in Japan. The national government has been playing an important role in furthering the understanding of climate science and its application. The national government provided direction through reports such as 'Direction of Climate Change Adaptation (Commission related to the Direction of Climate Change Adaptation, 2010)' by the MOE, the 'Wise Adaptation' report by the MOE, and the 'Climate Change Adaptation Report by Item' to

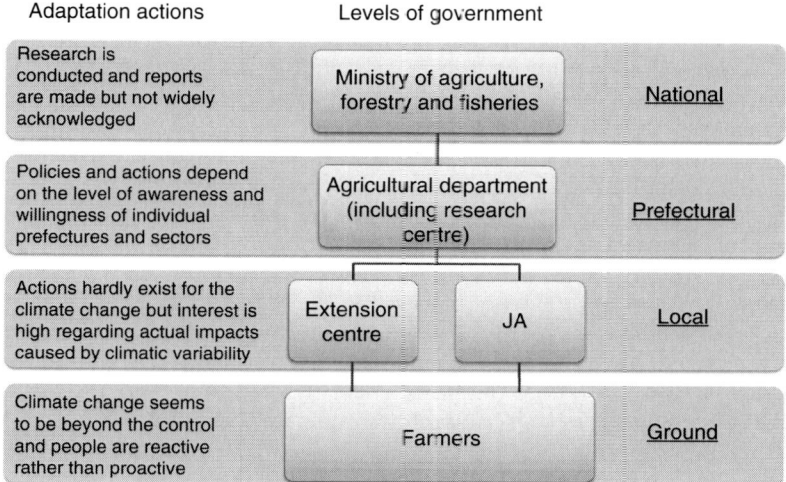

Figure 6.1 Status of CC adaptation actions at different levels of government with a focus on agriculture
Source: Authors' assessment of field surveys and consultations.

facilitate decision-making at the prefectural level (Committee on Impacts of Global Warming and Adaptation, 2008; MAFF, Japan, 2007). The national government also supports research in the field of CC forecast, impact and vulnerability assessment, and to a limited extent the identification of adaptation technologies and practices.

However, most of the current national initiatives on adaptation are limited since those initiatives are still at the stage of research. They are not widely acknowledged and are not at the stage of resource mobilisation for wider implementation. In addition, most of the existing sector-based adaptation projects are independently conducted by the different ministries in different periods (Table 6.1). They are not fully coordinated with other relevant sectors. Since the process of dialogue among various affected groups is required to deal with complex CC issues (Brown *et al.*, 2011), the current approach could easily cause fragmentation and unnecessary duplication.

At the local level the line departments such as agricultural extension and other collaborating institutions such as JA have not conducted any training programmes that are aimed at awareness generation or imparting specific skills for farmers to deal with long-term CC impacts. Although the MAFF has conducted CC impact assessments in collaboration with counterparts in each prefecture to facilitate technical assistance for agricultural production, it has not yet reached the local level. CC action hardly exists despite direct interest in local climatic impacts.

On the ground, communication channels between prefectural agricultural research centres and farmers appear to be highly dependent on JA due to the limited personnel within the prefectural extension departments. CC seems to be beyond the control of farmers. Everyday work is their priority, which makes them reactive to the impacts of CC. In fact, CC impact varies every year with forecasts affected by uncertainties, which makes it difficult for farmers to respond from a long-term perspective.

The context of adaptation is seen differently in Japan, where strategies to enhance adaptive capacity in the farm sector is dominated more by 'quality' concerns than 'yield' concerns, which is the priority in most developing countries. At the national level, initiatives have been established to deal with CC adaptation more in terms of promoting research on CC impact projections that could set the path for planned adaptation in the future. More concrete actions on the ground have been initiated by prefectures, including the development of heat-tolerant high-quality rice varieties, downscaling climate impact projections for agriculture, and chalky grain warning systems for maintaining high-quality rice. At the local level, people are found to be more concerned about dealing with existing climate variability than long-term CC.

The adaptation picture in Japan is emerging but fragmented, and more needs to be done. While well-informed approaches are something that may set Japan apart from other countries, the lack of a mechanism that coordinates and scales up local initiatives could result in the 'reinvention of the wheel' by other actors. Hence we propose a framework for horizontal and vertical coordination among ministries, sectors and players to 'scale up' the initiatives already taken up at the prefecture and local levels, and to make them interact with the government's plan and support (Howden *et al.*, 2007; Reid, Huq and Murray, 2010; Regmi, 2012). Such a coordination mechanism is possible at the national level. Since Japan already has relatively high adoption rates of advanced agrotechnology, the next step is to make available downscaled CC projections to all prefectures for initiating appropriate actions at the local level. In addition, Japan needs to address the changing socioeconomic conditions, including demographics and its implication for CC adaptation. One important lesson that could be learned from Japan, and that has direct relevance to the developing countries in the Asia-Pacific region, is that the local governments and institutions are strengthened to initiate adaptation actions even without significant support from the national level.

6. Limitations

This chapter is based on consultation meetings and interviews with various government institutions at the prefecture level. Though such an approach is not problematic, it could be difficult, as it is difficult for government officers to acknowledge the existing gap. In addition, the study has shown that technological interventions still play a major role in CC adaptation when compared with socioeconomic and community-driven approaches, which appear to be at the centre of CC adaptation initiatives in most developing countries. However, with the framework this research has followed, we cannot conclusively say that there are no community-based initiatives in Japan, but they are definitely not as visible as technological interventions.

7. Conclusion

Overall, although developed countries put less emphasis on the need for adaptation, this study revealed that the institutional system for adaptation in Japan needs to be strengthened. There are some actions emerging specifically at the local level of government and sectors but they still lack systematic approaches to efficiently deal with CC in terms of central-level coordination and strategising. Developed countries also need comprehensive policies similar to those of the developing countries. The following recommendations emerge from this study. First, there is a need for capacity-building of local functionaries for science-based adaptation. Topics such as downscaling of CC projections and impacts would be useful for local-level decision-making. Second, communication and coordination between different administrative levels needs to be improved, ideally with a greater role at the national level such that lessons learned across the prefectures can be shared effectively. Third, there is greater need for coordinated and integrated policies and programmes for avoiding duplication in different sectors and governments. Finally, there is a need for the diversification of adaptation strategies from technology orientation to social orientation.

Acknowledgements

We acknowledge the funding support received from the Asia Pacific Network for Global Change Research through project no. CRP2011-02NMY-Pereira and the S8 Project of Government of Japan. The authors are also grateful to several policy researchers, non-governmental organisations and policy-makers who attended the consultation meeting on 13 July 2010 in Japan.

Abbreviations

CC	climate change
GHGs	greenhouse gases
JA	Japan Agricultural Cooperatives
JMA	Japan Meteorology Agency
MAFF	MAFF (Japan)
MEXT	Ministry of Education, Culture, Sports, Science and Technology (Japan)

MLIT   Ministry of Land, Infrastructure and Transport
MOE   Ministry of the Environment (Japan)
OECD   Organisation for Economic Co-operation and
       Development

References

A. Agrawal (2010) 'Local Institutions and Adaptation to Climate Change', in R. Mearns and A. Norton (eds.) *Social Dimensions of Climate Change* (Washington DC: World Bank), p.173.

J. Ayers and S. Huq (2009) Community-Based Adaptation to Climate Change: An Update, Briefing of International Institute for Environment and Development, http://pubs.iied.org/17064IIED.html, date accessed 30 December 2012.

A. Brown, M. Gawith, K. Lonsdale and P. Pringle (2011) *Managing Adaptation: Linking Theory and Practice* (Oxford: United Kingdom Climate Impacts Programme).

Commission related to the Direction of Climate Change Adaptation (2010) *Direction of Climate Change Adaptation (Kikohendo tekio no hokosei)* (Tokyo: Ministry of the Environment, Japan).

Committee on Impacts of Global Warming and Adaptation (2008) *Wise Adaptation* (Tokyo: Ministry of the Environment, Japan).

Council of Food, Agriculture and Rural Area Policies (2006) The Adequate Role of National and Local Government on Land Improvement Planning *(Tochikairyo jigyo ni okeru kuni to chiho tono tekisetsu na yakuwari buntan)* http://www.maff .go.jp/j/nousin/sekkei/nn/n _sikumi/pdf/buntan.pdf, date accessed 18 November 2010.

Council of Science and Technology Policy (2010) Status of Climate Change Adaptation Activities *(Kikohendo tekiosaku ni kansuru torikumijokyo ni tsuite)*, http://www8.cao .go.jp/cstp/sonota/kikoutf/1kai/siryo2.pdf, date accessed August 2010.

J.D. Ford and L. Berrang-Ford (2011) 'Introduction', in J.D. Ford and L.B. Ford (eds.) *Climate Change Adaptation in Developed Nations: From Theory to Practice* (Dordrecht, Heidelberg, London, New York: Springer), pp. 3–20.

Global Warming Research Center for Agriculture and Fishery (2008) *2008–2010 Pilot Project on Production Area Structural Reform to Cope with Global Warming (Chikyu ondanka taio sanchi kozo kaikaku moderu jitsho jigyo)* (Miyazaki: Global Warming Research Centre for Agriculture and Fishery).

Hokkaido Agricultural Policy Department (2009) *Documents about Rice (Komeni kansuru shiryo).* http://www.pref.hokkaido.lg.jp/NR/ rdonlyres/D227E5FF-779B-4B0F-A390 -B46CD3BF89AC/0/ komenikansurusiryou2010.pdf, date accessed 18 November 2010.

S.M. Howden, J. Soussana, F.N. Tubiello, N. Chhetri, M. Dunlop and H. Meinke (2007) 'Adapting Agriculture to Climate Change', *Proceedings of the National Academy of Sciences of the United States of America*, 104, 50, 19691–19696.

S. Huq and H. Reid (2007) Community-Based Adaptation: A Vital Approach to the Threat Climate Changes Poses to the Poor, Briefing of International Institute for Environment and Development, http://pubs.iied.org/17005IIED.html, date accessed 30 December 2012.

IPCC (2001). *Climate Change 2001: Impacts, Adaptation and Vulnerability. IPCC Third Assessment Report* (Cambridge, New York, Oakleigh, Madrid, Cape Town: Cambridge University Press).

Japan Meteorological Agency (2012a) Weather Statistics Information Mean Value of Past Meteorological Data in Niigata (*Kisho Tokei Jyoho, Kakono Kishodeta Kensaku, Heinenchi, Niigata*), http://www.data.jma.go.jp/obd/stats/etrn/view/nml_sfc_ym .php?prec_no=54&prec_ch=%90V%8A%83%8C%A7&block_no=47604&block _ch=%90V%8A%83&year=&month=&day=&elm=normal&view=, date accessed 30 December 2012.

Japan Meteorological Agency (2012b) Weather Statistics Information Mean Value of Past Meteorological Data in Miyazaki (*Kisho Tokei Jyoho, Kakono Kishodeta Kensaku, Heinenchi, Miyazaki*), http://www.data.jma gc.jp/obd/stats/etrn/view/nml_sfc_ym .php?prec_no=87&block_no=47830&year=&month=&day=&view=, date accessed 30 December 2012.

MAFF, Japan (2007) *Climate Change Adaptation Report by Item (Hinmokubetsu chikyu ondanka tekiousaku report)* (Tokyo: MAFF, Japan).

MAFF, Japan (2008) *Comprehensive Strategy to Global Warming Provision by MAFF (Norinsuisansho chikyu ondanka taisaku sogo senryaku)* (Tokyo: MAFF, Japan).

MAFF, Japan (2009) Status of Planted Rice Variety in Fiscal Year 2009 *(Heisei 21 nendo suiko uruchimai hinsyubetsu sakutsuke jyokvo ni tsuite)*, http://www.syokuryo.maff.go .jp/archives/data/21kome-sakutsuke.pdf, date accessed 19 November 2010.

MAFF, Japan (2010a) Result of Rice Inspection *(Komeno kensa kekka)*, http://www .maff.go.jp/j/ soushoku/syoryu/kensa/kome/index.html, date accessed 2 September 2010.

MAFF, Japan (2010b) About MAFF Global Warming Task Force *(Nourinsuisansho ondanka taisaku honbu ni tsuite)*, http://www.maff.go.jp/j/press/kanbo/kankyo/ 091026.html, date accessed 24 September 2010.

Miyazaki Prefecture (2012) Third Miyazaki Prefecture Global Warming Countermeasure Implementation Plan *(Dai 3ki Miyazaki kencho chikyu ondanka taisaku jikko keikaku)*, http://www.pref.miyazaki.lg.jp/parts/000166955.pdf, date accessed 27 December 2012.

National Agricultural Research Centre for Tohoku Region (2006) *Annual Report 2006* (Iwate, Japan: National Agriculture Research Center for Tohoku Region).

Niigata Prefecture (2006) *Agriculture and Fishery Vision 2006 (Norin suisan vision 2006)*, (Niigata: Niigata Prefecture).

Niigata Prefecture (2009a) *Niigata Global Warming Regional Countermeasure Promotion plan 2009 (Niigataken chikyu ondanka taisaku chiiki suishin keikaku 2009)* (Niigata: Niigata Prefecture).

Niigata Prefecture (2009b) *Dream Developing Policy Plan (Yume okoshi seisaku plan)*, (Niigata: Niigata Prefecture).

OECD (2010) *OECD Environmental Performance Reviews: Japan 2010* (Paris: OECD Publishing).

K. Ouchi (2010) *Rice Prefecture Niigata Crying for Hot Summer (Komedokoro Niigata, mousho ni naku)* (Tokyo: Asahi Shinbun Press), at 9 October 2010.

Pacific Consultants Co LTD (2010) *Project Report of Reviewing and Supporting Climate Change Adaptation Vision (Kikohendo tekio shishin kento shien chosa itaku gyomu)* (Tokyo: Pacific Consultants Co LTD).

B. R. Regmi (2012) 'Revisiting Community-based Adaptation', *Forestry Nepal*, http:// www.forestrynepal.org/publications/article/5643, date accessed 31 December 2012.

H. Reid, S. Huq and L. Murray (2010) *Community Champions: Adapting to Climate Challenges* (London: International Institute for Environment and Development).

H. Seino (2008) *Global Warming and Agriculture (Chikyu Ondanka to Nougyou)* (Tokyo: Seizando Press).

K. Yamashita (2009) 'There is no Agricultural Reform Because of JA (Nokyo ga arukara nosei kaikaku ga dekinai)', Research Institute of Economy, Trade and Industry, http://www.rieti.go.jp/jp/papers/ contribution/yamashita/61.html, date accessed 18 August 2010.

M. Yokozawa and T. Iizumi (2009) 'Chapter 5 Rice Yield' in Project Team for Comprehensive Projection of Climate Change Impacts (ed.) S4 Second Report: Global Warming Impacts on Japan; Long-term Climate Stabilisation Levels and Impact Risk Assessment, http://www.nies.go.jp/s4_impact/pdf/S-4_report_2009eng .pdf, date accessed 27 December 2012.

7
How Adaptive Policies Are in Japan and Can Adaptive Policies Mean Effective Policies? Some Implications for Governing Climate Change Adaptation

S.V.R.K. Prabhakar, Misa Aoki and Reina Mashimo

1. Introduction

The Asia-Pacific region is one of the most climate change (CC) vulnerable regions in the world due to the relatively high proportion of its population depending on climate-sensitive sectors, dense population living in CC vulnerable geographical locations, and poor development of risk-governance systems. The national communications submitted by the developing countries to the United Nations Framework Convention on Climate Change (UNFCCC) showed gaps in its capacity, including research, to effectively cope with CC impacts (Kreft *et al.*, 2011). The need for enhanced adaptation research and policy-making capacity in developing Asia was recognised in a series of stakeholder consultations conducted by the Institute for Global Environmental Strategies (IGES) and the work was carried out at Universiti Kebangsaan Malaysia, M.S. Swaminathan Research Foundation and Vietnam Institute of Meteorology, Hydrology and Environment (Pereira *et al.*, 2011).

IGES consultations concluded that practical demonstrations on promising mainstreaming options, capacity strengthening and streamlining financial mechanisms are crucial to making further progress. Furthermore, many policy-makers called for identifying metrics or indicators to monitor the effectiveness of adaptation actions. Mainstreaming adaptation concerns in sectoral policy-making is relatively new and research on adaptation metrics is almost non-existent.

Even today, important policy decisions in the agricultural and water sectors are made and implemented without the consideration of projected impacts of CC (Srinivasan and Prabhakar, 2009). One of the most important barriers identified was the limited capacity of researchers in the region to provide adaptation policy-relevant information. For example, research on indicators for monitoring the effectiveness of adaptation options at different spatial scales is completely lacking. Networking and communication among researchers and policy-makers focusing on adaptation are also extremely limited.

In the absence of adaptation-specific information for decision-making, one of the schools of thought suggests that promoting basic ingredients of adaptive decision-making may help to support effective adaptation to CC (Peterson *et al.*, 1997; International Institute for Sustainable Development and the Energy and Resources Institute, 2006). According to this school of thought, promoting dynamic systems that can respond to known threats in a strategic manner reflects well the adaptive capacity of the system in question, and these systems are able to deal with CC and related uncertainties better than systems that are not 'dynamic' and 'adaptive' in nature.

Japan being a developed country and the signatory of the UNFCCC Kyoto Protocol, it has an obligation to reduce greenhouse gas (GHG) emissions by 6 per cent compared with the base year of 1990. At the national level, the emphasis appears to be more on the mitigation of GHGs than on CC adaptation (IGES, 2011). In addition to this there is an apparent understanding within Japan that its perceived threat from CC in agricultural and allied sectors could easily be managed. However, this does not mean that nothing has been done in CC adaptation. Several research programmes have been taken up within Japan to understand CC impacts and to implement actions on the ground (see Chapter 6 in this volume). These actions are dispersed across different ministries disguised under different names without being named as 'climate change adaptation'. This gives an impression that Japan is yet to go far in designing and implementing a clear CC adaptation policy in terms of being clearly stated in its relevant national and provincial policy documents. For this reason this study assumes an importance for understanding the adaptive nature of policy-making in Japan and its implications for CC adaptation there, and for those countries that consider the country as being a leader in the field of CC.

2. Adaptive policies, policy dynamics and their role in climate change adaptation

There are several definitions of the term 'policy' (Torjman, 2005). However, for the purpose of this chapter and from the viewpoint of public administration, a policy can be defined as a 'purposive course of action followed by an actor or a set of actors in dealing with a problem or a matter of concern' (Anderson, 1984, p. 3). There is a body of literature on why governments enact policies (Woll, 1974; Ingram and Smith, 1993; Considine, 2005; Torjman, 2005; Kay, 2006; Gerston, 2010). Most of these opinions converge to state that collective action should enable society to consume public goods and that a combination of several market failures affect the way in which public goods are produced, distributed and consumed (Weimer and Vining, 1992). Hence, the origin of the role of government in enacting policies is to enable equitable use of public resources, such as public goods.

On one hand, CC negatively impacts the developmental gains achieved by public (and private) interventions in past decades (Parry *et al.*, 2007), it impacts public goods (e.g. public infrastructure), resources (e.g. biodiversity and forests) and the well-being of individuals (e.g. livelihoods). On the other hand, CC would require public and private actions to mitigate GHG emissions and CC impacts. Therefore, CC is a public problem, requiring public solutions with collective action, and hence it is a subject of public policy (Dessler and Parson, 2010; International Institute for Sustainable Development, 2011).

CC is ridden with uncertainties in terms of future projections on the nature and degree of impacts (Schneider and Kuntz-Duriseti, 2002; Manning *et al.*, 2004), hindering credible and proactive actions, including policy interventions to mitigate the negative impacts. However, uncertainties should not be the reason for inaction (Maslin and Austin, 2012), and principles of adaptive management and adaptive policies should help in handling greater part of uncertainty (Peterson *et al.*, 1997; International Institute for Sustainable Development, 2011). The concepts of adaptive systems, adaptive management and policies hinge upon the fact that they help in developing alternative hypotheses, identifying gaps in knowledge and setting priorities (Peterson *et al.*, 1997).

Though the concept of adaptive policies is not new, the usage of this term in the context of CC adaptation can be traced to the International Institute for Sustainable Development's project entitled 'Designing Policies in a World of Uncertainty, Change and Surprise' (International Institute for Sustainable Development, 2011). However, the basic notion of a policy being dynamic dates back to several years before the beginning of the 2000s and has strong roots in a branch of policy science called policy dynamics (Baumgartner and Jones, 2002). This branch studies the feedback connections between the conditions and actors that are responsible for the development of a policy over a time period. According to this branch of policy science, policies can either remain unchanged over a period of time or change in a very predictable or unpredictable manner depending on the actors involved and the stimulus to which these actors respond. The evolution of this branch of policy science has strong roots in policy studies in the United States and benefits from the analysis of several decades of policy experience in that country.

Few similarities and contrasts can be drawn between the concept of adaptive policies and policy dynamics. The similarity between adaptive policies and policy dynamics is that both deal with how a policy evolves over a period of time and how they deal with the dynamic pressures that operate within a domain where a policy is made to operate. The concept of adaptive policies states that policies have to deal with both known and unknown conditions operating within the sphere of influence that they have, that they may lead to unknown and unintended consequences and probably may not be as effective as they are designed to be (International Institute for Sustainable Development and the Energy and Resources Institute, 2006). This, in the science of policy dynamics, is considered as positive and negative feedback processes that induce equilibrium and stability in the system (Baumgartner and Jones, 2002). Both concepts deal with the institutions that are involved in designing and implementing policies and how (that is the processes through which) policies are made. Hence, it can be concluded that a good understanding of policy dynamics can help the CC adaptation community well.

Understanding from both schools of thought – that is, policy dynamics and adaptive policies – seems to suggest that those policies and policy-making environments, including institutions and circumstances under which policies are made and implemented, that consider a broad range of conditions in designing and implementing policy solutions reflect better the ability for CC adaptation since such systems are able to deal with the uncertainties that are inherent in problems such as CC. This chapter aims to test the veracity of this understanding and its implications for CC adaptation.

For this study, Japan was chosen for three reasons: (i) In the international negotiations and the negotiation text (e.g. in the case of negotiations carried out under the UNFCCC) there is a consensus among many countries that developed countries have the capacity to adapt and help developing countries to adapt by transfer of technology and other related knowhow from their experience, (ii) Japan has been in the forefront in various aspects of environmental and CC policies, and (iii) Japan has serious concerns about food self-sufficiency and hence policy effectiveness in this is of paramount importance for the country.

Keeping the above background in view, the current research aims to examine whether all policies are characterised as adaptive would essentially lead to effective policies. Here, policy effectiveness relates to meeting the main objectives that these policies are intended to achieve. For example, several agricultural policies (see Table 7.1) in Japan have objectives of achieving food self-sufficiency and keeping the farming population within farming. We compared to what extent various amendments made to agricultural policies are able to achieve their objective as reflected in the published data.

3. Research methodology

This chapter is based on a Japanese case study entitled 'Strengthening capacity for policy research on mainstreaming adaptation to climate change in

Table 7.1 Major agricultural issues faced by Japan and major policy interventions addressing the issues during the past seven decades

Period	Major policies/acts/events	Driving issues*
1942–1960	Staple Food Control Act, Agricultural Cooperatives Act, Agricultural Improvement Promotion Act, Land Improvement Act, Agricultural Land Act, Act on Promotion of Agricultural Mechanisation, Act on Subsidies for Agricultural Improvement	Labour outflow into other industries, farmland dominance by landlords, reconstruction needed for subsistence farming framework, and decline in farming population in rural areas
1961–1972	Agricultural Basic Act, Forestry Basic Act, Amendment of Land Reform Act, Establishment of Japan Agricultural Cooperatives	Labour outflow into other industries, full-time farmers decrease, part-time farmers increase, soil natural capability decrease due to overusage of chemical fertilizers and pesticides, income disparity between rural and urban community, and farming population decline in rural areas
1973–1982	National Land Use Planning Act, Act on Agricultural Land, Agricultural Land Use Promotion Act; Act on Promotion of Improvement of Agricultural Management Foundation, Committee on National Rice Cultivators	Labour outflow into other industries, full-time farmers decrease, part-time farmers increase, income disparity between rural and urban community, Farming population decline in rural areas, and environmental pollution issues
1985–1992	Agreement on Multipolar Pattern National Land Formation, General Agreement on Tariffs and Trade, Ministry of Agriculture Forestry and Fisheries of Japan announced 'A New Way to Food, Agriculture and Rural Policy'	Income disparity between rural and urban community, full-time farmers decrease, part-time farmers increase, cultivated land abandonment, and farming population decline in rural areas

Table 7.1   (Continued)

Period	Major policies/acts/events	Driving issues*
1993–2001	Establishment of environmentally sound agriculture implementation headquarters, Act on Stabilization of Supply, Demand and Prices of Staple Food; Repeal of Staple Food Control Act; Minimum Access System of Rice; Act on Special Measures on Incentive Loan Program for Youths to Become Farmers, New Rice Policy, Agricultural Policy Reform, Food, Agriculture and Rural Areas Basic Act, Act on Promoting Sustainable Agricultural Production Practices, Hilly and Mountainous Area Direct Payment System	Change in farmland usage (farmland liquidation), decrease in full-time farmers, increase in part-time farmers, increasing abandonment of cultivated land, ageing of farming community
2002–2010	Management Policy for Promoting Structural Reform of Agriculture Report, New Rice Policy, Amendment of Food Control Act, Restriction of Genetically Modified Crops by Local Governments, Measures and Policies for the Improvement of Conservation of Rural Land; Comprehensive Strategy on Countermeasures Against Global Warming, Trans-Pacific Partnership, Income Compensation System for Individual Rice Farming Households	Change in farmland usage (farmland liquidation), excess production of rice, decrease in full-time farmers, increase in part-time farmers, increasing abandonment of cultivated land, ageing of farming community, food security, increasing need for adaptation to CC

Source: Adapted from Ohara and Soda (1994), p.168.

agricultural and water sectors' funded by the Asia Pacific Network for Global Change Research (Project Number CRP2010-02NMY-Pereira). As part of this project a consultation meeting with various stakeholders (total 28 partici-pants) involved in policy research and government in agricultural and allied sectors was conducted on 28 June 2011 at the Japan Press Centre Building,

Tokyo, to understand how dynamic policies and institutions in Japan are formulating and implementing various policies related to agriculture and natural resource management. The participants were selected based on their expertise in agricultural policy processes in Japan. The participants discussed the policy environment in agricultural and allied sectors in Japan, how dynamic it is, and reasons behind the effectiveness of policies. The specific subjects discussed were the historical analysis of agricultural policies in Japan, the declining number of farmers in Japan and the evolution of related policies, historical analysis of interventions to deal with floods and droughts in Japan, and fiscal policy support for dealing with agricultural and natural resource-management issues in Japan.

Considering the theoretical background presented in the previous section, the framework used for assessing the policy effectiveness and adaptiveness of policies in this study include asking a set of questions: (i) when the policies were introduced to address the perceived problem, (ii) how frequently the policies were amended to address changing circumstances, (iii) how effective the policies introduced are and (iv) how the effectiveness is related to when and how frequently policies were introduced. These questions have also formed the guiding sections for discussion in this chapter. The policy effectiveness is judged by comparing the policy objectives with the trend in certain indicators, such as the area under agricultural land and the size of the farming population.

A questionnaire survey was conducted to get a consensus on the issue of adaptive policies and to identify important issues in the agricultural sector in Japan and the related policies introduced. While the consultation meeting was used to understand the overall agricultural policies and issues in Japan, the questionnaire survey helped us to obtain ranked opinions on the policies. The questions asked for information about respondents, identification of important issues in agriculture in Japan, important policies introduced, and opinion about selected policies for each issue ranked high by the respondent. The respondents included PhD students and experts in agricultural policy in Japan considering their knowledge and expertise on policy issues related to agriculture and natural resource management. Eight responses were obtained from 30 questionnaires sent by the time of the initial drafting of this chapter. Since this is a pilot survey the results should be viewed as provisional. The results corroborate the discussions in the consultation meetings.

4. Findings and discussion

4.1 When policies were introduced?

In order to answer this question, historical analysis of various agricultural and allied policies in Japan was conducted from the available literature and the findings are presented in this section (see Table 7.1). The purpose was to identify a policy as 'dynamic' if it undergoes continuous change over the years as a result of external pressures operating on agricultural and allied sectors.

Table 7.1 presents a list of important driving forces that operated during various phases of agricultural policy development and policies that have been implemented in Japan in the past seven decades (modified and substantially updated from Ohara and Soda, 1994). Agricultural policy development in Japan can be broadly divided into six time periods – that is, post-war reconstruction period (1940s–1950s), Post-Agricultural Basic Act (1960s), low economic growth period (1970s to early 1980s), globalisation period (mid-1980s to early 1990s), structural reform of agricultural and rural policies period (most of the 1990s) and realignment of agricultural and rural policies to global trends (most of the 2000s).

Driving forces for policies introduced during these periods vary greatly. During the post-war reconstruction period (Table 7.1) the driving forces for policies were labour flow, the dominance of landlords, reconstruction of the economy and the decline in farming population in rural areas impacting food self-sufficiency. The government had to address these issues early on by introducing policies such as the Staple Food Control Act (1942), the Agricultural Cooperatives Act (1947), the Agricultural Land Act (1952), the Act for Promotion of Mechanisation (1953) and the New Rural Construction Act (1956). All of these acts very much correspond to the issues identified during that period. The same follows for most of the driving forces and policies mentioned in Table 7.1.

From this table we can conclude that agricultural policy environment in Japan can be characterised as either 'reactive' or 'adaptive' since the government is able to continuously introduce new policies and amend old ones (refer to our definition of adaptive policies earlier in this chapter). It is reactive for the reason that mostly the policies were made in response to emerging issues, but mostly well within a decade, within which these policies were identified and implemented with a reasonable period of identifying the issues by the policy-formulating institutions and stakeholders. However, this conclusion should be read with caution since there is no way for this research to identify 'when' a particular issue or driving force has come into existence since most agricultural policy issues have no clear beginning and end point but rather seamlessly emerge over time. Nevertheless, from this review it can be broadly concluded that agricultural policies in Japan were made in immediate response to the issue once it came to the notice of the policy-makers in the country. This addresses the question of how soon a policy was made and brought into effect in Japan.

4.2   How frequently were policies amended?

To answer the question of how frequently policies have changed over the period (amended or repealed), following the changing circumstances or driving forces, the number of amendments and repeals some major policies have undergone were tabulated (Table 7.2).

It is clear from the table that some policies have undergone very frequent changes, as often as every year during their implementation (e.g. Agricultural Cooperatives Act, Agricultural Land Act and Food, Agricultural and Rural Areas Basic Act), while others have remained more or less the same (e.g. Agricultural Improvement Promotion Act and Act on Subsidies for Agricultural Improvement).

From Table 7.2 the following conclusions can be drawn: (i) the high frequency of changes may have to do with the importance of the issues that these policies address, (ii) frequent changes in governments, possible lack of consensus within government and institutions responsible for their formulation and implementation, inability of earlier versions of policies to stem the issue and (iii) lack of clear understanding among institutions and governments on how to address the problem. However, what these also show is the willingness of governments to tackle the issues with continuous efforts at policy level seeking a correct solution. By this, governments and institutions appear dynamic in nature and hence have the ability to adapt to changing external pressures affecting policies.

4.3 How effective are the policies?

While the question of how soon a policy was introduced and how frequently it was modified to keep abreast with changing circumstance is important, even more important is that the policy delivers the intended outcomes (i.e. meeting its objective). To identify the effectiveness of policies, they were overlaid on the time series diagrams of various indicators which reflect the effectiveness of a policy for a better visual representation.

4.3.1 *Number of farmers*

Declining numbers of farmers has been a major cause of concern for Japan as this is leading to heavy reliance on imported food, thus burdening the national economy (Namiki, 2007). Various specific policies and amendments were introduced to control the outflow of farmers from agricultural to non-agricultural sectors and to increase new recruits into the farming sector.

Figure 7.1 depicts the major policies introduced and their effectiveness on the trend in the number of farmers (full and part time). It is clear from the figure that the policies introduced over the years have not been able to control the outflow of farmers as reflected by the continuous decline in number of farmers in the country.

4.3.2 *Decline in farmland*

A factor that is closely associated with the declining number of farmers is the associated decline in acreage of farmland. Figure 7.2 shows the trend of

Table 7.2 Amendments in major agriculture and related policies in Japan

S. N	Policy/act	Number of amendments	Period when the amendments were carried out	Frequency (changes per year)
1	Staple Food Control Act	27	1943–1994	0.5
2	Agriculture Cooperatives Act	83	1948–2010	1.3
3	Agricultural Improvement Promotion Act	16	1950–2004	0.3
4	Land Improvement Act	55	1951–2011	0.9
5	Agricultural Land Act	66	1953–2010	1.2
6	Act on Promotion of Agricultural Mechanisation	13	1962–2006	0.3
7	Act on Subsidies for Agricultural Improvement	16	1961–2010	0.3
8	Agricultural policy	3	1978–1999	0.1
9	Act on Promotion of Improvement of Agricultural Management Infrastructure	19	1989–2010	0.9
10	Act on Stabilisation of Supply, Demand and Prices of Staple Food	9	2000–2010	0.9
11	Act on Special Measures Concerning Incentive Loan Program for Youths to Become Farmers	11	1995–2010	0.7
12	Food, Agriculture and Rural Areas Basic Act	10	2000–2010	1.0
13	Act on Promoting the Introduction of Sustainable Agricultural Production Practices	3	2002–2010	0.4
14	Act on Special Measures for Promotion of Independence for Underpopulated Areas	9*	2000–2011	0.8
15	Policy for Delivering Subsidies to the Farmers for Stabilisation of Agriculture	1	2009	0.0

*with another amendment scheduled in 2016.
Source: the authors.

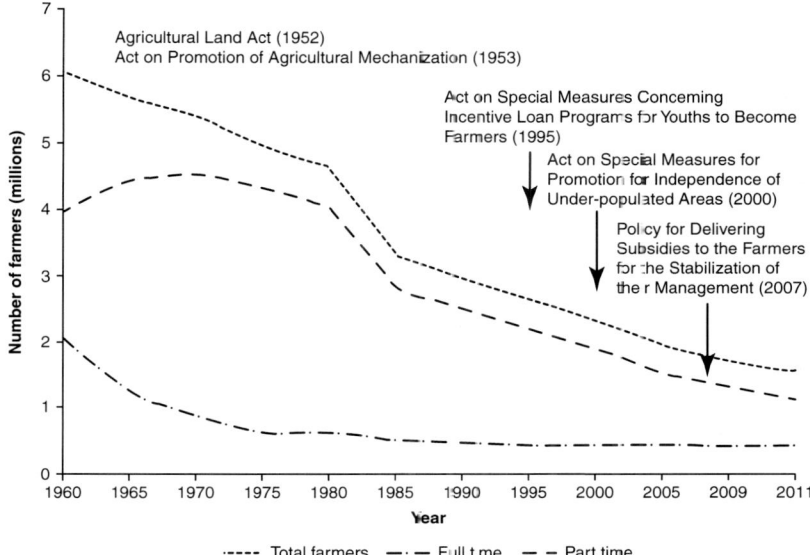

Figure 7.1 Trend in the number of farmers in Japan over the past five decades and various policies introduced to stem the decline in their number

Source: Adapted from the Ministry of Agriculture Forestry and Fisheries of Japan (2011a and 2011b).

total population, agricultural production, usage rate of cultivated land and number of farmers. As in the earlier case, several policies were introduced to control the change in land use from agricultural to non-agricultural purposes, though some initial leverage was applied for the deliberate movement of land to non-agricultural purposes for promoting industrialisation during the early years of economic growth in Japan. However, such policy support for land conversion has slowly been withdrawn in recent years (Kazuhito, 2008). The main policy introduced to control the land-use change from agricultural to non-agricultural was the Amendment of Land Reform Act (1970) and other related policies. Figure 7.2 also shows that none of these policies could stem the continuous decline of farmland over time. Please refer to the limitations part of this chapter for more explanation of this conclusion.

From the above examples of trends in farming population and land-use changes it is clear that related policies have failed to stem the trend. More interestingly, these are the policies that have undergone most amendments since they were introduced (e.g. the Agriculture Land Act has undergone 66 amendments, Table 7.2). It can be concluded from these observations that the indicators such as 'how soon policies were introduced' and 'how frequent policies were amended' may not necessarily lead to effectiveness in policy outcomes.

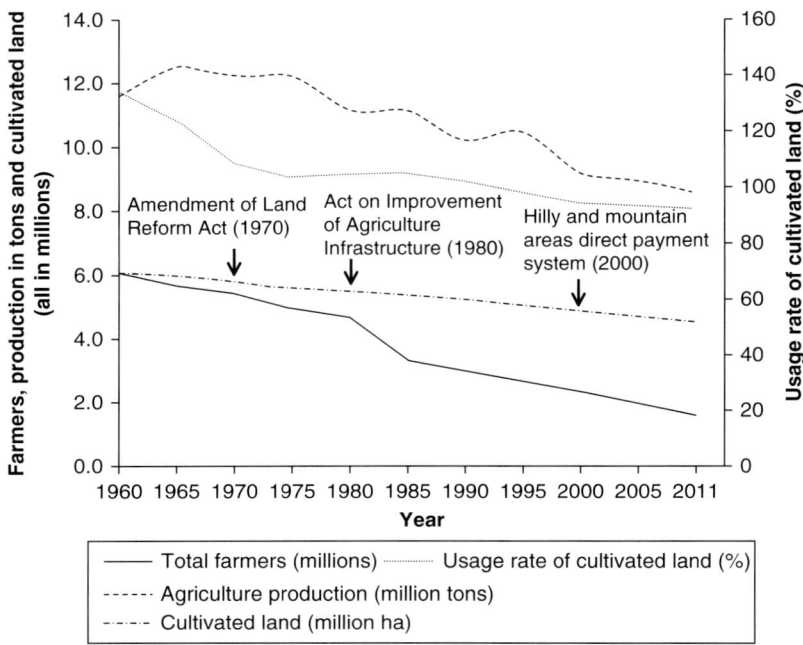

Figure 7.2   Land-use changes and various policies introduced to control land-use change
Source: Adapted from the Ministry of Agriculture Forestry and Fisheries of Japan (2011a and 2011c).

4.4   Results of pilot survey on adaptive policies

Most respondents indicated the decline in number of farmers as a main policy issue for agriculture in Japan (38 per cent) and they opined that the Agriculture Basic Law or any law that supports farmers and group farming is an important policy intervention for Japan. As the second most important policy issue, most respondents ranked declining global competitiveness of Japanese agricultural produce followed by increasing income gap between rural and urban areas in Japan (see Table 7.3 for the responses).

Respondents were asked to rate specific policies for their timeliness, adaptiveness, effectiveness and strategy on a scale of 1 to 5 where 1 is least timely and 5 is most timely. Those who said that the declining number of farmers is an important policy issue in Japan have rated the related policies as least timely, least to moderately effective (which is corroborated by the Figure 7.1), least to moderately adaptive and least to moderately strategic in nature.

Overall the respondents were not satisfied with the effectiveness of policies introduced in Japan. This very much corroborates the discussion in

Table 7.3 Important issues identified and policies suggested by the respondents in the first round of the Delphi Survey

Rank category	Important issue hindering agriculture in Japan	Important policies for overcoming these issues
First	Declining number of farmers	• Agricultural Basic Law • Support for new farmers and group farming
Second	Declining global competitiveness of Japanese agriculture	• Protecting domestic agriculture • Promoting minimum access policies • Promoting industrialisation
Third	Increasing income gap between rural and urban areas	• Subsidies for mountainous areas • Compensate farmer income

$n = 8$, pilot survey.
Source: the authors.

Section 4.3 wherein the introduction of different policies did not lead to positive changes in the trend of the number of farmers and land used for agricultural purposes.

5. Limitations

By nature, due to reasons not clear to us, agricultural issues may remain 'under the carpet' or 'invisible' until they surface after crossing a threshold. Identification of this period from literature is often difficult and was outside the scope of this research. Hence, we could not pinpoint the exact year when a particular policy problem came into existence for the purposes of assessing the timeliness of introducing policies. The policy effectiveness was assessed by comparing the trends in certain indicators such as the size of the farming population and the area under agriculture. Though several policies were introduced to stem the declining trend in these indicators, one could see that these continued unabated (Figures 7.1 and 7.2). Though we concluded that this is a clear indication of policy failure, the observed trends could also have happened due to forces outside the purview of the agricultural sector. For example, globalisation, lucrative jobs in technology and the service sector, which provide a better income and working conditions, have a much stronger driving force than the solutions offered by the introduced policies to keep people in farming. Taking all of these outside forces into consideration would further strengthen the study.

However, it is still safe to conclude that agricultural policies failed to take into account what is happening outside agricultural sector and hence can be concluded as reason behind policy failure. This stresses the need for comprehensiveness in understanding and the need for policies to have broad reaching impact for policies to be effective.

6. Conclusion

One of the important criteria for assessing the readiness of a country to adapt to CC has been reported as its ability to formulate and implement policies in an adaptive manner which can be evaluated in terms of how soon policies are implemented and how frequently they undergo changes to reflect the changing circumstances. This chapter presents the results of a pilot survey that corroborates the findings from the literature review and the consultation meeting conducted on this subject.

From the preliminary assessment presented in this chapter, it is clear that though countries like Japan have a good history of formulating and implementing several policies to address perceived issues in agriculture, the mere assessment of these policies in terms of how soon they were introduced and how often they were modified doesn't explain the policy effectiveness. The effectiveness of a policy would go beyond these indicators/criteria presented in this chapter. The additional criteria for the effectiveness of policies could be whether they are designed based on the right stimuli, the correct perceptions of policy-makers of these stimuli, and if the policy is based on the right information. In addition, the evaluation of these policies should be done based on their outcome and should not be limited to indicators such as timeliness, which could be misleading, as clearly shown in this chapter. This has major implications for the community engaged in CC adaptation since this community needs to take decisions often based on limited information. Hence, providing policy-relevant information that is timely is crucial for effective policies.

Acknowledgements

We are grateful for the valuable input from the consultation workshop with various policy researchers, non-governmental organisations and policy-makers organised on 28 June 2011 in Tokyo, Japan. We also acknowledge the funding support received from the Asia Pacific Network for Global Change Research through project no. CRP2010-02NMY-Pereira and the S8 Project of Government of Japan.

List of abbreviations

CC	climate change
GHG	greenhouse gas
IGES	Institute for Global Environmental Strategies
UNFCCC	United Nations Framework Convention on Climate Change

References

J. E. Anderson (1984) *Public Policymaking: An Introduction*, 2nd ed. (Boston: Houghton Mifflin), p. 342.

F.R. Baumgartner and B.D. Jones (2002) *Theoretical Beginnings' in Policy Dynamics* (Chicago: University of Chicago Press), p. 360.

M. Considine (2005) *Making Public Policy* (Cambridge, United Kingdom: Polity Press), p. 262.

A. Dessler and E.A. Parson (2010) *The Science and Politics of Global Climate Change: A Guide to the Debate* (Cambridge, United Kingdom: Cambridge University Press), p. 211.

L.N. Gerston (2010) *Public Policy Making: Process and Principles* (United States of America: M.E.S harpe, Inc.), p. 166.

H. Ingram and S.R. Smith (1993) *Public Policy for Democracy* (Washington, D.C., United States of America: The Brookings Institution), p. 275.

IGES (2011) Proceedings of Consultation Meeting on Adaptive Policies and Measuring Mainstreaming Climate Change Adaptation into Institutional Processes: Some Experiences from Japan, International Forum for Sustainable Asia and the Pacific, Yokohama, Japan. 12–13 July, 2010.

International Institute for Sustainable Development (2011) Adaptive Policy Making, International Institute for Sustainable Development, http://www.iisd.org/climate/vulnerability/policy.asp, date accessed 15 October 2011.

International Institute for Sustainable Development and The Energy and Resources Institute (2006) *Designing Policies in a World of Uncertainty, Change, and Surprise: Adaptive Policy-making for Agriculture and Water Resources in the Face of Climate Change* (Manitoba, Canada: International Institute for Sustainable Development and The Energy and Resources Institute).

A. Kay (2006) *The Dynamics of Public Policy: Theory and Evidence* (Massachusetts, United States of America: Edward Elgar Publishing Limited), p. 149.

Y. Kazuhito (2008) *The Perilous Decline of Japanese Agriculture* (Tokyo, Japan: The Tokyo Foundation), p. 10.

S. Kreft, A.O. Kalonga and S. Harmeling (2011) *National Adaptation Plans: Towards Effective Guidelines and Modalities* (Bonn, Germany: Germanwatch and World Wide Fund for Nature), p. 28.

M. Manning, M. Petit, D. Easterling, J. Murphy, A. Patwardhan, H. Rogner, R. Swart, and G. Yohe (2004) *Intergovernmental Panel on Climate Change Workshop on Describing Scientific Uncertainties in Climate Change to Support Analysis of Risk of and of Options* (Maynooth, Ireland: National University of Ireland), p. 138.

M. Maslin and P. Austin (2012) 'Climate Models at Their Limit?', *Nature*, 486, 183–184.

Ministry of Agriculture Forestry and Fisheries of Japan (2011a) *Census of Agriculture and Forestry* (1960–2010) (Tokyo, Japan: Ministry of Agriculture, Forestry and Fisheries). http://www.maff.go.jp/j/tokei/ census/afc/index.html, date accessed 27 February 2013.

Ministry of Agriculture Forestry and Fisheries of Japan (2011b) *Survey of Agricultural Structure* (Tokyo, Japan: Ministry of Agriculture, Forestry and Fisheries), http://www.maff.go.jp/j/tokei/ kouhyou/noukou/, date accessed 27 February 2013.

Ministry of Agriculture Forestry and Fisheries of Japan (2011c) *Statics of Farmland* (Tokyo, Japan: Ministry of Agriculture, Forestry and Fisheries), http://www.maff.go.jp/j/tokei/kouhyou/ sakumotu/menseki/ index.html, date accessed 27 February 2013.

M. Namiki (2007) 'Active Agricultural Population in Post-war Japan', *The Developing Economies*, 7, 2, 158–169.

K. Ohara and O. Soda (1994) 'The Development of Agriculture and Agricultural Policy and the Change of Views on Farming and Rural Society After World War II', *Sanbon Taihaibutsu Shigen Kino Murasaki*, 12, 167–181.

M.L. Parry, O.F. Canziani, J.P. Palutikof, P.J. van der Linden and C.E. Hanson (eds.) (2007) *Climate Change 2007: Impacts, Adaptation and Vulnerability. Contribution of Working Group II to the Fourth Assessment Report of the Intergovernmental Panel on Climate Change* (Cambridge, United Kingdom: Cambridge University Press), p. 976.

J.J. Pereira, J. M. Pulhin, S.V.R.K. Prabhakar, R. Shaw and R.R. Krishnamurthy (2011) 'Climate Change Adaptation: Perspectives in the Asia Pacific', *Asian Journal of Environment and Disaster Management*, 2, 4, 361–494.

G. Peterson, G.A.D. Leo, J.J. Hellmann, M.A. Janssen, A. Kinzig, J.R. Malcolm, K.L. O'Brien, S.E. Pope, D.S. Rothman, E. Shevliakova, and R.R.T. Tinch (1997) Uncertainty, Climate Change, and Adaptive Management, *Conservation Ecology* [online], 1, 2, 4, http://www.consecol.org/vol1/iss2/art4/, date accessed 10 February 2013.

S.H. Schneider and K. Kuntz-Duriseti (2002) 'Uncertainty and Climate Change Policy', In S.H. Schneider, A. Rosencranz, and J.O. Niles (eds.) *Climate Change Policy: A Survey* (Washington, D.C.: Island Press), pp. 53–87.

A. Srinivasan and S.V.R.K. Prabhakar (2009) *Measures of Adaptation to Climatic Change and Variability (Adaptation metrics)* (Hayama, Japan: The World Bank and Institute for Global Environmental Strategies), p.120.

S. Torjman (2005) *What Is Policy?* (Ottawa, Canada: Caledon Institute of Social Policy), p. 20.

P. Woll (1974) *Public Policy* (Boston, United States of America: University Press of America), p. 265.

D. L. Weimer and A.R. Vining (1992) *Policy Analysis: Concepts and Practice* (Englewood Cliffs, USA: Prentice Hall), p. 424.

8
Management of Climate-Induced National Security: Paradigm Shift from Geopolitics to Carbon Politics

Md Shafiqul Islam

1. Introduction

On an official visit by the Australian parliamentary secretary for the Pacific Islands to Bangladesh's climate vulnerable island, Char Kukri Mukri, on 12 November 2011, the secretary commented that the island is the ground zero for climate change (CC) (the author himself was present at the address ceremony). The term 'ground zero' used by the honourable secretary signifies the impending danger of CC not only to this island but to the entire South Asian region.

CC has the potential to pose a serious threat to the national security of the states that are affected by its adverse implications. It can act as a threat multiplier for instability in some of the most volatile regions in the world and can present significant security challenges to nations. Various research reports have already forecast the cascading effects of unmonitored CC on a range of security problems that could lead to dire global consequences. The United Nations (UN) Security Council has also recognised the impacts of CC as threats to international peace and security. The scientific assessment of the consequences of CC also paints a gloomy picture. Major vulnerabilities induced by climatic hazards include a rise in sea level, higher temperatures, greater intensity of severe natural disasters resulting in human displacement, shortages of drinking water, health hazards, loss of agricultural land and food scarcity (Mahmudul, 2010). As mentioned earlier, these have the potential to trigger instability at the national and regional levels. This new potential threat to national security is beyond human intentionality to harm a country and it needs pragmatic and innovative approaches to address these issues. The traditional security apparatus and its strategies need new focuses and methodologies to cope with this pervasive threat to national security. In this chapter an attempt is made to recapitulate the changing paradigm of climate-induced threats to national security and the management of threats and changes.

The wide ranging vulnerabilities of the adverse effects of CC are a well discussed issue. There are a large number of research works and publications on these subjects. But the evolving trends and new implications of CC are continuously drawing the attention of all groups of stakeholders. Similarly, national security is another much discussed and well-researched subject that has prompted a huge number of publications, research and ongoing discussions. It is also an evolving subject which continuously draws the attention of state functionaries. However, there is a distinct link between these two issues which is rarely discussed or highlighted in the public domain. Unless extensive research is done or attention is drawn to this, the state apparatus will not be prepared to face the challenge. Against this backdrop, this chapter has been undertaken to attempt to bridge the research gap on this important issue.

2. Research objectives

The objectives of this chapter are to (i) examine the threats posed to national security by the adverse effects of CC and (ii) explore innovative approaches required for the mitigation of such threats and challenges in the context of the South Asian region.

3. Research method

The contextual framework of this chapter is based on secondary published sources and empirical research literature of co-related topics. It has been mentioned in the background that very few empirical research studies have been carried out on the management of climate-induced national security. So most of the information and references have been drawn from ongoing discussions by experts in seminars/workshops and related publications. It may be noted that I was involved in national security affairs throughout my long professional career. All of these sources helped in generating the ideas expressed here for the management of climate-induced national security.

4. National security in the context of the rapidly changing globalised world

Until recently the notion of national security was based on external threats from other states and from regimes associated with those states. The focus of such national security was geopolitical and military centric and all other dimensions of citizen's security were relegated to the background (Karim, 2006). Post-Cold War security concerns and global democratic practices broaden the focus of national security. The new focus was expanded to include peoples' security (e.g. health, education, welfare, basic needs, diseases, hunger and human rights), the environment and socioeconomic development. Thus a citizen-centric security discourse started occupying the central stage (Kabir, 2000). But the mechanism of threat perception and methodology to cope with such threats did not change much. Against this backdrop, after many devastating natural calamities, CC and its impact on national security gradually came to the forefront. It became evident that climate-centric disasters could cause more devastation than actual geopolitical state-centric aggression (Kabir, 1989).

5. The impact of climate change: An intertwined fabric of national security

5.1 Climate change: An overview

For most countries, CC can be extremely detrimental to the economy, the environment, national development and people's security. Issues are more likely to happen due to the increased frequency and intensity of natural disasters (Mahmudul, 2010) such as sea-level rise, increases in atmospheric temperature, drought, wildfire, tidal surge and so on, as well as inefficient management of the impacts. The 2006 US National Security Strategy noted that deadly pandemics or severe natural disasters can produce weapons of mass destruction effects (Busby, 2007). It further noted that megadisasters, such as floods, hurricanes, earthquakes or tsunamis, could swiftly kill and endanger a large number of people in the same way as an armed attack, and might overwhelm the response capability of the state. Such events could generate security threats, like the breaking down of social order, large-scale migration and so on. The impacts of CC are multidimensional and multispectral. It can simultaneously affect all aspects of state affairs and human habitation (interstate relations, economy, development, food production, infrastructure, human mortality, etc.). It is interesting to note that in the short term, for a few northern developed countries, CC may be perceived as advantageous due to the opening of northern sea routes. It may also gradually uncover the polar land mass, creating opportunities for enhanced mining, farming and fishing. Yet, it raises a security concern due to conflicting ownership claims and state rights. It also poses the risk of more disease, new forms of virus and the deformation of certain animal species. Some of the phenomena and their impacts on climate-induced national security are briefly discussed below.

5.2 Natural disaster

The natural disaster is not a new phenomenon but its intensity and severity are on the rise, probably due to the impact of CC. According to the fourth assessment report of the Intergovernmental Panel on Climate Change

(IPCC, 2007), the number of natural disasters may double during the next 10–15 years. It concluded that some weather extremes will become more frequent, more widespread and more intense during the 21st century. The recent statistics on natural disasters such as cyclones, floods, drought and wildfires are predicted to be linked to CC and they have devastating effects on the economy, development and people's security. For example, over the past ten years, 3,852 disasters killed more than 780,000 people, affected more than 2 billion people and cost a minimum of US$960 billion (Muniruzzaman, 2011). However, in the case of the recent events, such as cyclone Sandy (United States, 2012), hurricanes Katrina (United States, 2006), Ivan (United States, 2004), Andrew (United States, 1992), cyclones Sidr (Bangladesh, 2007), Nargis (Myanmar) and Ayla (Bangladesh, 2009), CC cannot be linked directly to these disasters but they do provide a visual image of the devastating effects of CC.

5.3 Sea-level rise

The 2007 IPCC report indicated a temperature rise of 1.1–6.4 °C around the world by 2100. Another report of the International Institute for Environment and Development found that a tenth of the world's population – approximately 634 million – live in the coastal areas of the world that lie between 0 m and 10 m above sea level. Most of these people will be affected and some of them have to be relocated due to an increased frequency of disaster and sea-level rises. Dr James Hansen, a former employee of National Aeronautics and Space Administration's (NASA's) Goddard Institute for Space, presented a gloomy outlook in his book titled '*Storms of my grandchildren; the truth about the coming climate catastrophe and our last chance to save humanity* (Hasan, 2011). He mentioned that a temperature rise of even 2 °C or more would make the earth as warm as it was 3 million years ago. Such warmth can cause sea levels to be about 25 m higher than they are today (Hasan, 2011). Even this relatively modest change would have serious consequence (like the loss of habitat, food scarcity and the loss of housing and agricultural land) for the vast population of coastal areas, creating an unmanageable number of climate refugees. This has the potential to create humanitarian disaster on an unprecedented scale.

5.4 Loss of agricultural productivity

One of the effects of CC is likely to be the loss of prime agricultural land and sources of irrigation. This could result in food insecurity, hunger and disease, and could also severely impact the national economy of the affected countries. The UN Food and Agricultural Organisation (FAO) has warned that the unprecedented drought across the major food-exporting countries in 2012 (e.g. the United States, Ukraine and Russia) has reduced the world's grain reserve to a dangerously low level. Due to drought, flood, cyclone and other natural disasters, the world consumed more than it has produced in

2012, pushing the world's reserve to its lowest level since 1974. If there is another such major natural disaster in 2013, it could create a worldwide food shortage, leading to social unrest and even major violence. The FAO report further mentioned that 870 million people across the globe are already malnourished and any further depletion of food reserves could result in the breakdown of the global food supply-chain system. Leading environmentalists predict that the world's climate is no longer reliable and poses a more serious threat than armed aggression, and that this threat could result in food insecurity, which could spark widespread riots, bringing down governments and disrupting civil order. This climate-induced food insecurity by causing famine could lead to internal political uprising in some countries and might even threaten their national security as a result.

5.5 Scarcity of water and tension across international rivers

The importance of water for human survival is well known and its scarcity due to CC and other human-related factors has led to tensions among neighbouring countries in Asia and the Middle East. Due to the rapid melting of snow in the Himalayas, more water is likely to be available for a short period across the international rivers of this region. However, the state-centric myopic management approach of international rivers might fuel the existing water conflict between neighbours, posing a heightened threat to national security.

5.6 Impact on national security

In a globalised world, national security extends well beyond the protection of territorial integrity against belligerent states (Saber, 2008). Like armed attacks, some of the effects of CC could swiftly kill, displace or endanger a large number of people (Joshua, 2007). At the same time it could cause catastrophic disruption to infrastructure and public health, and create civic unrest beyond the management capabilities of most of the would-be-affected developing countries. Thus the human security of a country could be seriously affected. CC even poses an existential threat for some of the small island countries. The IPCC report of 2007 predicts that even developed countries such as the United States may not escape the dangers of CC, especially in coastal areas (IPCC, 2007). A NASA simulation report indicates that a 40 cm sea-level rise by 2050 combined with a Category 3 hurricane could inundate much of New York City (Busby, 2007). The reality of the effects of such a disaster was observed when cyclone Sandy hit the New York coast, causing unprecedented damage, including the flooding of railway tunnels for the first time in the recorded history of the city. A more direct effect of CC on national security could occur through the destruction of military establishments. This could have a catastrophic effect on national security because the situation could easily be exploited by hostile forces jeopardising the country's national security. Moreover, the weakening of its defence capabilities could further exhaust its disaster-response capacity. This would have a direct impact on the population who would be isolated by the severe natural disaster. In the post-disaster scenario only defence forces can reach inaccessible areas with their equipment and organisational support. So if the disaster-response capacity of the military of a country were affected it would ultimately affect the human security of that state.

5.7   Operation 'Sea Angel': The case study of Bangladesh

In 1991, southeastern Bangladesh was hit by a severe cyclone. This major destructive force affected Chittagong, the second largest city in the country. Bangladesh's only major sea port, its only major naval base and second largest airbase is also located within the perimeter of the city. The cyclone caused widespread damage, including to military aircraft and naval vessels. The port, air base and naval base became non-operational during this period, reducing the state's ability to effectively execute rescue operations. The cyclone also significantly reduced the defence capability of the air force and navy. In such a scenario, miraculous help came from the US 7th Fleet which was returning to the East Pacific from the Gulf War under General Stack Pole. The general immediately launched a humanitarian operation called Sea Angel, deploying helicopters, air lifts, hovercraft and landing vessels delivering food, water and medicine to the survivors in inaccessible areas. The incident clearly highlights how natural disasters can affect national defence capability, but at the same time it helps us to realise that the capacity gap can be filled through international cooperation.

6.   The dawn of climate diplomacy and carbon politics

6.1   Climate diplomacy

The climate diplomacy concept is a new one. It is issue based, whereas traditional ongoing diplomacy is ideologically based. Thus on the issue of CC, groups and subgroups have emerged based on common interests and potential gain. One such group is the big polluter group, which leverages its emission power like Morgenthau's power politics (Hans, 2001). It is against any binding carbon emission cut and looks for lacunas to adhere to its policy. For example, during the signing of the Kyoto Protocol, the United States opted out on the plea that China and India should also commit to cutting carbon emissions. But China and India did not agree, citing millions of their population at the underdeveloped level. Ultimately everyone remains the victims of excess emission of greenhouse gases (GHGs) and global warming. During the 17th Conference of the Parties (COP 17) in Durban in December 2011, the European Union wanted China, the biggest polluter on earth, to commit to deeper cuts in emissions, but

China came up with its own arguments and pushed the conference to the point of failure (Rahman, 2011). The COP 17 talks took on the characteristics of the Cold War era ping-pong diplomacy, while Canada, Japan and Russia (umbrella group members) shifted to a wait-and-see strategy from their earlier proactive and selfless exemplary roles. Similarly the G-77, a longstanding group, was divided over myopic national interest, losing its collective bargaining power and collective leadership role. In this case the active role of the African group had a sabotaging role in the capacity of least developed countries (LDC). Small Island States (including Asian countries such as the Maldives, Mauritius and Madagascar) that are in the front line of CC formed their own group to mitigate their sufferings and pleaded for their own agenda. By this division the impacting countries from both sides continuously came up with their own agenda and the most vulnerable ones were relegated to the sidelines. Even after all of this the Durban conference came out with a declaration that was acceptable to all. However, Greenpeace lamented that the Durban declaration was a victory for polluters over people because the allowed level of carbon emissions of the big polluters will endanger the life of the millions across Asia and Africa.

6.2 Carbon politics

The political discussion and decisions taken at different international conferences to ameliorate the effect of carbon emissions is best described by Nobel laureate Desmond Tutu as 'adaptation apartheid' because this is characterised by polluting industrial countries spending huge amounts of money enhancing their adaptive capacity while the LDCs, which do not contribute to pollution yet are worst affected by CC, are grossly underfunded in their adaptation efforts. Observing the deep-rooted national priority regarding global danger at Durban, COP 17, the conference chairman of the UN Framework Convention on Climate Change, the South African foreign minister, Maite Nkoana Mashabane, said that we should realise that the Durban package may not be perfect but we should not let the perfect become the enemy of the good and the possible (Reuters, 2011). When the world and its 7 billion people are endangered by the curse of development, still the nations are playing the tricks of carbon politics. The politicians are failing to realise that it is a global problem which requires a global solution rather than the self-destructive national approach (Munim, 2011). The most vulnerable countries are also playing the cards of carbon politics to gain from their sufferings. Unless all carbon politicians are engaged in a soul-searching exercise, it will be very difficult to save the earth from the curse of carbon and dangerous development. A carbon-induced CC bomb is ticking away in the poor regions of Asia and it is a major political challenge to diffuse it among global economic stress and carbon politics.

7. Management approach of climate-induced national security

7.1 Policy

The currently practised defence policy is based on geopolitical considerations and human-induced armed attack. It is state centric and the focus is on the external threat from belligerent states or regimes. This needs reorientation and expansion to incorporate CC impacts. Such changes in the defence policy will not only insulate the country itself from the adverse effects of CC but will also address the concern of strategic partners. At the same time, the policy will open up the vista of new partners based on environmental vulnerability and its adverse effects. Some of the policy issues requiring immediate inclusion are discussed below.

7.1.1 Constant learning model focusing on climate-change impacts on national security

The impact of CC on national security is a comparatively new phenomenon to the defence forces. Most of the technical and general issues are not in the defence curriculum. At the same time the focus is not there. So the inclusion of all aspects of CC into the security studies curriculum might not be enough for learning the details. Special effort, a long-term learning strategy and a constant focus could result in the effective awareness and knowledge-building of the defence forces on the impacts of CC on national security.

7.1.2 Developing military-to-military contacts on climate-change issues

Current military-to-military cooperation is based on a common threat perception and its linkage to strategic alliance, complimentary defence development and resource optimisation, as in the North Atlantic Treaty Organisation. The security concerns arising out of CC issues are not included in this. It needs reorientation and the contacts of and cooperation between the different military groups to be developed based on the common grounds of CC effects. Also, the military groups that have the potential to help in implementing CC mitigation and adaptation strategies should be included in the cooperation framework.

7.1.3 Research and development on climate-change mitigation and adaptation issues

Traditionally, advanced military groups have played a leading role in inventing technology and methodology that ultimately augmented fire power, mobility, lethality and counteracting the capabilities of the opposition. Defence research and development seldom focused on natural phenomena and their adaptation and mitigation mechanism. The preservation of nature, the mitigation of CC and response to disaster and GHGs were seldom in the military's research domain. But the adverse effect of CC on national security requires the defence researchers and the authorities to set objectives that will help to achieve the national goals for combating the adverse effects of CC and its preventive and adaptive strategies regarding national security.

7.1.4   Application of coercive force for the implementation of global policies and rule

For a long time military groups have been engaged in achieving national defence and economic interests and for UN mandates. We are familiar with military, economic and trade blockades, safe areas, neutral zones and no-fly zone concepts. Yet, it is unheard of and beyond comprehension that the military can be used for the protection of the environment and climate. To avert climate-induced security effects the military's coercive forces may be required to enforce global policies on GHG reduction, a green economy and carbon-intensive product reduction on the states or regimes that do not comply with such policies. Thus more pragmatic and innovative military action is needed to address such issues. New action like a carbon blockade or applying coercive force against carbon rouge nations is gradually gaining attention. In this scenario the difficulty for the military is the broad spectrum and canvas of these issues. It is well justified to incorporate these issues into an implementation framework for the military. However, the climate and environmental issues are of a regional and global nature, and it is difficult for the military to justify any action within the framework of the national interest and objectives. To act in such scenarios, new alliances, such as the Climate Vulnerable Group, whose members share a common interest, are of paramount importance. These alliances should be without geographical proximity, in other words geopolitical consideration. Climate and environmental issues will be the prime considerations that will guide the government to focus such military cooperation frameworks. Another new development for the military will be to include in the grouping regional bodies, states, companies, civil societies, non-governmental organisations and rights-based organisations with which the military has seldom interacted in the past.

7.1.5   Develop climate-change expertise and knowledge-sharing framework

As mentioned earlier, the climate- and carbon-related undertakings are new frontiers for the military. Thus it will not be easy for it to play the necessary role without gaining sufficient knowledge about the various aspects of climate issues. Some issues are of a purely scientific nature while others are economic, legal, trade and awareness related. For the military there is a need to develop expertise in each field. Unless this expertise is acquired, it will not be in a position to work side by side with other military and civil bodies in an effective manner. It was mentioned earlier that a climate-induced threat to national security is of a regional and global nature. So the expertise needs to be shared within a regional framework, such as the Association of South East Asian Nation (ASEAN) or the South Asian Association for Regional Cooperation (SAARC), or by forming a special group of countries that are likely to face a similar level of threat to national security emanating from the adverse effects of CC.

7.2 Macrolevel action plan

For the implementation of policies to mitigate the threats arising from CC, targeted action plans are of prime importance. These could range from developing appropriate strategies for risk reduction to the policy for adaptation and mitigation issues. Some elements of the risk-reduction strategies could be to help in developing legally binding laws and bodies like that of the Law of Sea Treaty to monitor and implement globally accepted adaptation and mitigation policies. Through research and countries' own or regional expertise, attempts should be taken to produce more precise estimates of CC effects, encourage climate-suitable building code and identify safe living areas that are less vulnerable to CC effects. These risk-reduction strategies will be more beneficial to the state than improvement of the already built infrastructure, settlement or recovery/relocation strategy after a disaster. Some of the specific macrolevel action plans are considered below.

7.2.1 *Additional budget for the military for climate-change-related activities*

Currently a global opinion prevails regarding defence spending cuts for enhanced development, particularly in the context of the LDCs. But some of these counties, especially Asian countries such as Bangladesh, the Maldives, Sri Lanka, India and Nepal, are most vulnerable to the adverse effect of CC. There, the military is one of the most important organs to respond to the problem. So to achieve climate resiliency, adaptability and reverse CC activities, the defence forces of these countries must play important role. At the national, regional and global levels, defence forces could be employed for the implementation of (i) carbon control and emission policy, preventing deforestation activities, (ii) carbon-trade and carbon-compensation policy and (iii) green economy and green environment policy. These are new frontiers for the military which demand new capacity and the formation of new alliances. To develop the necessary defence force capabilities to face the pervasive effects of CC on national security, enhanced funding for the military is a prerequisite and is justifiable.

7.2.2 *Increased learning modules among regional military groups*

It was mentioned earlier that the CC impact on national security is not state centric. It is more of a regional and global phenomenon. Under such circumstances, for the military to play any role at the macrolevel requires frequent interaction and knowledge-sharing among regional military groups. When the enhanced peacekeeping operation under the UN started after the First Gulf War, Asian military groups established a peacekeeping academy/institute and shared knowledge and experience to build each others' capacities at the regional level. Similar organised efforts should be undertaken to face the challenge to national security emanating from CC. These could be achieved by holding regular discussions, brainstorming sessions, conferences, workshops and so on between the military groups on CC-related national security issues. These could ultimately help to establish a regional and global military framework to collectively deal with the CC menace.

8. Limitations

The main intent of this chapter is to examine the vulnerabilities of CC in the context of national security. A broader framework has to be adopted in the absence of well-documented country-specific analysis. The main limitation of the chapter is the non-availability of quotable data relating to a particular country. Structured defence discussion and analysis are still highly restricted areas, particularly in the South Asian context. In such a scenario, more reference had to be made to Western states, though climate vulnerabilities are more prominent in South Asian countries.

9. Conclusion

The security threat of CC is perceivable, massive and of regional magnitude. It is far more severe than the state-centric, human-induced and geopolitics-based national security parameters. Policy, diplomacy and strategy based on national power and interest is not very effective in this context. New considerations, such as the stabilisation of GHGs, emission control, tariffs on the excessive use of carbon, a green economy, carbon trading and so on should be the focus of the military during the present era. The military should understand that in the coming years, security dimensions are going to be changed from geopolitics to carbon politics, from human-based intentionality to the mercy of nature, and from a state-centric perspective to a regional and global perspective. Such a dramatic shift of security parameters requires a pragmatic and innovative approach to save nature and, at the same time, the state. A state-centric security apparatus is not very effective in an era when CC can pose the biggest threat to national security, particularly to some of the most vulnerable countries of Asia. To determine the new roles and regional working modalities for the military groups in the face of such a security threat, a detailed study should be pursued under regional bodies such as ASEAN and SAARC.

Abbreviations

ASEAN   Association of South East Asian Nations
CC   climate change
COP   Conference of the Parties
GHG   greenhouse gas
IPCC   Intergovernmental Panel on Climate Change
LDCs   least developed countries
NASA   National Aeronautics and Space Administration
SAARC   South Asian Association for Regional Cooperation
UN   United Nations

References

Z. Hasan (2011) 'Has Global Warming Doomed Bangladesh?' *The Daily Star*, 11 October 2011, Dhaka, Bangladesh.

B.W. Joshua (2007) *Climate Change and National Security, An Agenda for Action*, CSR No 32 (New York: Council on Foreign Relations).

M.H. Kabir (2000) *National Security of Bangladesh in Twenty First Century* (Dhaka: Academic Press and Publishers Limited).

M.G. Kabir (1989) 'Environmental Challenges and the Security of Bangladesh', *BIISS Journal*, 10, 1.

M.A. Karim (2006) *Contemporary Security Issues in the Asia-Pacific and Bangladesh* (Dhaka: Academic Press and Publishers Library), p. 37.

H.J. Morgenthau (2001) *Politics among Nations. The Struggle for Power and Peace* (New Delhi: Kalyani Publishers).

I. Mahmudul (2010) *Progress Report on Disaster Risk Reduction and Climate Change Education, Research and Training in Bangladesh, Comprehensive Disaster Management Program (CDMP), A Program of the Government of Bangladesh* (Dhaka: Government of Bangladesh).

A.N.M. Muniruzzaman (2011) 'Climate Change: Threat to International Peace and Security', *The Daily Star*, August 11, 2011, Dhaka, Bangladesh.

R. Munim (2011) 'Disaster Risk Loom Large', *The Daily Star*, 10 October 2011, Dhaka, Bangladesh.

A. Rahman (2011) 'Durban Conference: Most Successful Failure!', *The Daily Star*, 11 December 2011, Dhaka, Bangladesh.

Reuters (2011) 'Draft Climate Accord Emerges, but Problems Remain', *The Daily Star*, 11 December 2011, Dhaka, Bangladesh.

M. Saber (2008) 'National Security of Bangladesh: Challenges and Options', *NDC Journal*, 7, 1.

9
Deconstructing Debate on the National Action Plan on Climate Change at the State Level: A Case Study of Meghalaya State, India

Ashok Kumar Singha, Suvra Majumdar, Abhik Saha and Somnath Hazra

1. Introduction

The general debate about environmental issues in India centres on the fact that the threats of climate change (CC) have affected the majority of the poor disproportionately. As per the Planning Commission, Government of India, 12th Plan document (2012), the projected changes by 2100 include high regional variability in rainfall (15–40 per cent) with extreme weather conditions and more dry days. Other projected changes include increased temperature in land areas in northern India, a relatively longer winter and post-monsoon seasons, and an increase in annual average temperature by 3–6 °C. There are also reported findings that suggest significant changes in glaciers in the Himalayan ecosystem and resulting water-balance changes in the river basins, especially in the northeastern region and the Indo-Gangetic plains as per the Indian Network for Climate Change Assessment (Ministry of Environment and Forest, 2010). The main effects of CC will reduce agricultural productivity and fish production in the coastal areas (Ministry of Environment and Forest, 2010). Given that most of the poor participate in the agricultural workforce, they are likely to suffer the most. According to a report by the Mckinsey Global Institute (2010), India's urban population is projected to increase from 340 million in 2008 to 590 million in 2030 due to migration from rural areas. These migrants will suffer heavily due to CC mainly as a result of resource congestion. The city boundaries are generally planned for a specific amount of water, habitat, energy use, waste, transport and green cover. Continued in-migration will upset this balance and increase the overexploitation of these resources, leading to greater emissions. There-fore, whether in rural or urban areas, there is likely to be a great deal of heterogeneity in the vulnerability of India to CC.

Guiteras (2007) used district-panel data to estimate the impact of weather shocks on gross productivity in agriculture. As per his estimate, in the medium-term CC scenario, crop yields will decline by 4.5–9.0 per cent, which is consistent with the estimate by Sanghi, Mendelsohn and Dinar (1998), who used data from 271 districts and found that 1 °C warm-ing would reduce the net farm revenue by 9.0 per cent for all categories of farmers. The other debate is centred on the common but differential responsibility of the nations to mitigate and adapt to CC, and to have equal rights to the atmosphere. Since the developed region of the world has overexploited resources to clutter the carbon space, leading to grow-ing emissions, any developing nation taking climate positive action needs to be incentivised by developed countries so that the equity is maintained (Tol, 2001).

Considering all models (TERI-Poznan, TERI-MoEF – both take into account the CO_2 emissions from energy and industry), Irade (CO_2 from energy, indus-try, household and government consumption only), NCAER models (CO_2 and N_2O emissions from energy and industry only), by 2030, India, despite its high economic growth rate of 8.5 per cent per annum, is likely to have a scenario where the per capita emissions will still be below the 2005 global average of 4.22 tons (Ghosh, 2010). India links energy security to its commit-ment to the Millennium Development Goals. Also, it would not compromise its growth objective and poverty-reduction effort by agreeing to any signifi-cant emission-reduction target without international support on technology and finance. It is also India's position that it would not agree to any regime that would result in welfare loss through carbon tax (at a moderate tax of US\$20/ton; a loss of US\$1,194 billion at 2005 prices) (Ghosh, 2010). If India sticks to its voluntary emission-reduction targets in a higher growth regime; under two different scenarios, the maximum welfare loss in 2050 in terms of equivalent variation is estimated at 6.35 per cent (in the carbon-tax sce-nario) and at 5.16 per cent in the common differential convergence scenario. (Pradhan and Ghosh, 2012).

2. Research objectives

The objective of this chapter is to highlight the process followed in engaging the subnational (state-level) stakeholders in CC debate and derive a series of priority actions through a prioritisation matrix and develop a climate bud-get using multiple criteria. This process starts with understanding the state vulnerability, mapping the national action plan (sectors) and facilitating the stakeholders to brainstorm and suggest various actions relating to adaptation and mitigation.

3. Research methods

In this chapter we follow a case-based approach where we illustrate the process of formulation of the State Action Plan on Climate Change (SAPCC). The methodology has been developed based on our experience in the formulation of such plans in several states. The case of the state of Meghalaya in India has been used here as an example. We were involved first hand in the review of climate data, collection of primary data and interviews with working group members, key informants, both on a one-to-one basis. The state has notified the working-group members and 100 per cent of them were covered. Then meetings were held in each of the five agroclimatic regions of the state and from each zone about 50 participants, including leaders from community-based organisations, women, farmers, members of industry associations, forest-dependent communities and so on. We organised consultation workshops with these stakeholders. This chapter describes the key processes involved in developing a state action plan on CC in a participative manner which engaged various groups of stakeholders in a CC debate.

The steps conducted in this research study were as follows: (i) developing an institutional framework to start the debate on CC at the state level to develop the SAPCC; (ii) linking state departments/sectors with national missions based on the mission objectives; (iii) identifying state vulnerability due to CC using a vulnerability-assessment framework; (iv) engaging stakeholders in CC-related deliberation in a structured way to identify actions that can help in adapting to CC or in mitigation or both; (v) prioritising actions; and (vi) developing a climate budget using multiple criteria which we developed in Orissa state and which are now used in many other states.

4. National Action Plan on Climate Change

India's first National Action Plan on Climate Change (NAPCC) outlining existing and future policies and programmes addressing climate mitigation and adaptation was released on 30 June 2008. It identifies eight core 'national missions' running through 2017. Embedded in the departmental architecture, the eight missions also have measurable targets and clearly identified nodal departments to drive the missions, as summarised in Table 9.1.

4.1 Rationale for a state-level action plan on climate change

The missions stated in Table 9.1 converge to address issues relating to adaptation as well as mitigation actions to contain the adverse impact of CC.

The idea of a subnational action plan emerged as it is grounded locally and is expected to (i) reflect high ownership, (ii) create better awareness linking

Table 9.1 Missions under National Action Plan on Climate Change

Mission	Objective	Responsible Ministry
1 *National Solar Mission*	• To increase the share of solar energy in the overall energy mix increasing 1,000 MW PV per year and deploy at least 1000 MW of solar thermal power and overall at least 20,000 MW solar power by 2020	Ministry of New and Renewable Energy
2 *National Mission for Enhanced Energy Efficiency*	• To save about 10,000 MW by the end of the 11th Five-Year Plan in 2012 through energy efficiency measures.	Bureau of Energy Efficiency, Ministry of Power
3 *National Mission for Sustainable Habitat*	• To undertake energy efficiency measures in buildings (both commercial and residential), promote waste management and attempt modal shift through sustainable public transport	Ministry of Urban Development
4 *National Water Mission*	• To undertake water conservation measures through integrated water resource management and drive it through water policy and regulation to optimise water use efficiency by 20%	Ministry of Water Resources
5 *National Mission for Sustaining the Himalayan Ecosystem*	• To empower local communities especially Panchayats to play a greater role in managing Himalayan eco-systems	Ministry of Science and Technology
6 *National Mission for a Green India*	• To increase land area under forest cover from 23 per cent to 33 per cent using measures like compensatory afforestation programme and joint forest management to have about 6 million hectares of afforestation of degraded forest lands by the end of 12th Plan	Ministry of Environment and Forests

| 7 | *National Mission for Sustainable Agriculture* | • To make Indian agriculture climate resilient using drought proofing, conservation, crop diversification and protection techniques apart from risk transfer through insurance | Ministry of Agriculture |
| 8 | *National Mission on Strategic Knowledge for Climate Change* | • To enhance knowledge and capability for adaptation and mitigation through collaborative research and development with global community | Ministry of Science & Technology |

Source: Compiled from Government of India (2012). *National Action Plan for Climate Change*, accessed from www.india.gov.in/allimpfrms/alldocs/15651.doc 17 October 2012.

experiences of climate-linked issues to corrective actions, (iii) allow better preparedness and (iv) set strategic priorities at the subnational level.

These priorities would enable the leaders in the states to make plans, organise resources and save on the long-term costs associated with CC mitigation and adaptation.

Considering the varied geography of India, state priorities and issues vary significantly. India occupies 2.4 per cent of the world's land area and has diverse physiographic features, such as the Himalayan areas, coastal areas, the northern plains, the peninsular plateau and islands. It is also home to more than a billion people. The majority of Indians depend on monsoon-linked rain-fed agriculture (more than two-thirds of the population). It is also endowed with a varied climate, biodiversity and ecological regions – for example, forest cover is about 6,755,000 sq. km, which is under acute anthropogenic pressure (Ministry of Environment and Forest, 2010).

While macrolevel models (IPCC 4th assessment report) demonstrate the threats of CC, the states are required to take action to mitigate such threats. Due to this high degree of diversity, the first approach was required to carry out detailed scoping at the state level using a strategic environmental analysis and vulnerability assessment tools (Boehringer, Fischer and Rosendahl, 2010). In some states, even the state-level emission inventory was developed. The second important aspect of detailing an action plan on CC at the state level is to create a link to the developmental planning process and resource transfer. This requires an institutional framework that takes into consideration the available resources under a 'business as usual scenario' and additional funding requirements for CC adaptation and mitigation efforts. Subsequent sections in this chapter deal with the case of the Meghalaya state action plan on CC, including the underlying methodology, tools and processes.

5. Case study on the development of the Meghalaya State Action Plan on Climate Change

The state action plan on climate change in Meghalaya is based on a logical framework that links three distinct strands of the NAPCC – namely, (i) a robust institutional framework and intrinsic linkage to the NAPCC, (ii) a vulnerability framework to set the context and (iii) tools and processes to identify key priorities to address CC issues in the state. This case study is based on the draft report that has been prepared and is available to the public.

5.1 Institutional framework

5.1.1 Institutional architecture at the national level

The institutional architecture to guide national policy on CC action planning is through the Ministry of Environment and Forest. The national communication channels are divided thematically – namely, (i) inventory, (ii) vulnerability assessment and (iii) uncertainty reduction. Joint secretary CC coordinates with states for the preparation of state-specific action plan on CC.

5.1.2 Institutional arrangement at the state level

In the case of Meghalaya, the Meghalaya State Council on Climate Change and Sustainable Development (MSCC&SD) coordinates the state action plan for the assessment, adaptation and mitigation of CC. The composition of the MSCC&SD is discussed as follows.

The chief minister is the chairperson and other members are:

- minister, forest and environment
- minister, agriculture and irrigation
- minister, water resources
- minister, soil and water conservation
- minister, science and technology
- minister, power and non-conventional energy resources
- chairman, State Planning Board
- chief secretary Meghalaya
- vice chancellor NEHU, Shillong
- additional chief secretary, planning
- additional chief secretary, finance
- chief executive members, district councils
- principal secretary/commissioner and secretary, planning (is also the secretary, Forest) convenor.

This combination ensures interdepartmental convergence in CC action planning (Government of Meghalaya, 2010).

Necessary government notification of this council was the key architecture to provide guidance for CC action planning in the state. GiZ (German Development Cooperation) provided the technical support through CTRAN. Further, a steering committee was formed to facilitate the whole process. The members were the secretaries/heads of the relevant departments who functioned as nodal officers. Each nodal officer also acted as a convenor of a group to list the technical actions that the group deemed fit to address the issue of CC in the state. The following are the key milestones of the Climate Change Action Planning Process (CCAP) in Meghalaya:

- CC action plan strategic dialogue (November 2010);
- deliberation on the formulation process (November 2010);
- inception meeting with key officers and stakeholders (December 2010);
- notification of the institutional framework for the preparation of the CCAP (January 2011);
- establishment of nine working groups for the preparation of the plan (January 2011);
- preliminary draft of the CCAP (March 2011);
- stakeholder consultation on the draft (June 2011);
- revision based on the stakeholder input and finalisation (August 2011);
- submission of draft CCAP to the government of India (September 2011) (Government of Meghalaya, 2012).

5.2 Matching the key sectors in the state with the national missions

The first step was to identify the top issues on CC as per the national priority (see Table 9.1).

There is a logic why sectors identified in the SAPCC have to be mapped against the national missions. Buys *et al.* (2007) argued that there is a clear distinction between source and impact vulnerability. While in their paper, they did a cross-country comparison on source and impact vulnerability at the country level, this is manifested in cross-state comparison. The negotiations articulated at the national level take into account the national targets and commitments. Unless similar commitments and factors constitute the debate at the state level, then the mission objectives and national commitments may not be achieved. States may have very unique state-specific climate stress. In such circumstances they have the option to get the necessary support from nationally prioritised missions through federal resource transfer. If this source of funds is not available or inadequate, they can fund state-specific adaptation and mitigation actions through the state budget (Table 9.2).

CC debate is also linked to resource transfer and burden-sharing. There are states in India who are more polluting or have larger quantity of emission

Table 9.2   Meghalaya state priorities aligned with the national mission

National mission	Key Departments in Meghalaya	Key issues
Sustainable Agriculture	Agriculture, Horticulture, Fishery	Flood-resistant varieties, methane management
Green India	Forest, Tourism	Forestry, Bio-diversity, ecotourism
Energy Efficiency	Energy, Industry	Energy efficiency in Micro Small and Medium Enterprises (MSME) sector, food-processing, mineral-based industries
Solar mission	Energy	Reduction of carbon foot print
Sustainable habitat	Urban development, Public Works Department (PWD), housing	Storm-water management, energy-efficient green building
Strategic Knowledge for Climate Change	Planning, Forest, Finance	Addressing cross cutting issues, developing a knowledge centre on climate change in the state
Himalayan Ecosystem	North Eastern Hill University (NEHU)	Academic research, glacial flow, delta management, local variability
Water	Water, Agriculture	Water management

Source: Government of Meghalaya (2012). *State Action Plan on Climate Change [draft]*, (Shillong: Department of Planning), p. 9.

than others. At the federal level, there is an option to incentivise clean states or provide for resources to more vulnerable ones from the better off states. Therefore, attempt has been made to map the national missions to state level vulnerable sectors primarily as a resource transfer strategy.

5.3   Vulnerability analysis to identify state-specific priorities

Vulnerability and adaptive capacities are diverse and vary from state to state and even from location to location within a state based on several sectoral and cross-sectoral parameters (Füssel, 2010). Sectoral parameters include key sectors of the state's economy and cross-sectoral factors include (i) poverty, (ii) inequality and social discrimination over property rights, (iii) access to resources, (iv) social attrition/migration and (v) unequal and unsustainable competition for scarce natural resources (Thomas and Twyman, 2005). They are also a function of the adaptive capacity and the exposure that the inhabitants have to the climate. In other words, the people who are exposed to more climate-stressed situations, such as flood, drought or waterlogging, are likely to be more vulnerable if they have a low adaptive capacity. If they have a better adaptive capacity, the vulnerability will be comparatively less under the same level of exposure.

Vulnerability analysis is critical for development and planning, and in some sense a paradigm shift from the way in which the planners should design policies and programmes in a CC lens.

The assessment included three distinct segments:

- biophysical factors: trends in degradation, forest cover, biodiversity and so on due to anthropogenic activity;
- socioeconomic factors: social groups like Scheduled Castes and Scheduled Tribes (SC–ST) that have high natural-resource dependency;
- climate sensitivity: trends and impacts of precipitation and temperature, and incidence of extreme weather events.

Regarding state-level CC impacts, descaled global and various national models (e.g. a combination of HadCM3 and PRECIS models is known as the HadRM3 model) are used to analyse trends and outcomes of various climatic parameters (see http://www.metoffice.gov.uk; Figure 9.1).

The data on district-wise projected increases in annual average temperature (°C) for the period 2021–2050 (A1B SRES scenario) compared with the baseline (1975) projected by the HadRM3 model shows the following. The western parts of the state are projected to experience a greater increase in temperature when compared with the eastern parts of the state but the variability in the increase in temperature is not great, with the

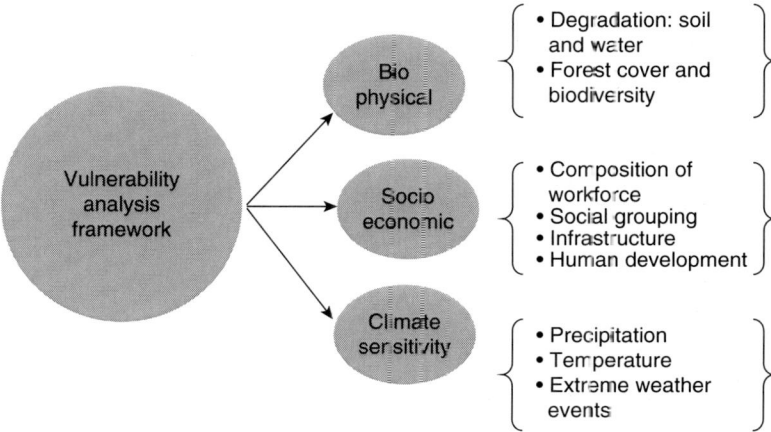

Figure 9.1 Vulnerability assessment framework
Source: Compiled from government of Meghalaya (2012). *State Action Plan on Climate Change [draft]* (Shillong: Department of Planning), p. 10.

biggest increase being 1.8 °C and the increase in the East Khasi Hills being 1.6 °C (Kumar and Parikh, 2001). Monsoon- and water-related vulnerability is widely seen in many areas of the state, especially those where there is a prevalence of monoculture, which suffer the most. While in the national discourse, Meghalaya is one of those states where there is always above-average rainfall, the actual distribution shows that the variability in rainfall has created havoc in the agriculture production system as well as biodiversity. This is one example where common knowledge at the national level may not address state-level vulnerability.

However, state vulnerability is more due to social grouping and the presence of more than 100 tribal groups. These groups have a biomass and natural resource intensive lifestyle and are more vulnerable to CC. Forestry is an important resource in Meghalaya accounting for 77.23 per cent of its geographical area, and a large section of the tribal population is dependent on it. Traditional shifting (*jhum*) cultivation is practised by the people, which impacts forest conservation. About 49 per cent of the state population are below the poverty line, and the state income is below the national average. High poverty and lower human development is attributed to the inherent climatic and social constraints mentioned above. The disaggregated data at the state level shows that the developments, wherever visible, are largely urban centric. This, in turn, puts a lot of emphasis on resource congestion and planning related to waste management, energy efficiency and water-use efficiency.

5.4   Sectoral strategy for climate change adaptation and mitigation

The working groups and experts, which included officers from technical departments, academicians, civil society groups and grassroots organisations, brainstormed on various issues relating to the sector based on the data and projected impact. A long list of actions relating to adaptation and mitigation were generated. Tools of prioritisation were used to shortlist at least ten high priority actions in each sector.

Based on multi-criteria analysis tool, rooted in the vulnerability of the state due to CC, an initial long list of more than 400 actions were discussed. Many actions not related to CC were avoided. From 103 shortlisted CC linked actions, 70 actions were identified as key priorities. The total budget for the priority actions is estimated to be about US$1,400 million (Indian Rupees (INR) 6,297 crore). This budgeting exercise is a rough estimate based on the norms for various investment projects. For research and capacity-building projects, budgets are estimated as per the activities. (Figure 9.2).

Multicriteria analysis is a method for designing integrated adaptation and mitigation issues taking cognizance of the policy trade-offs. The assumption is that this investment will enhance the adaptive capacity of the state and

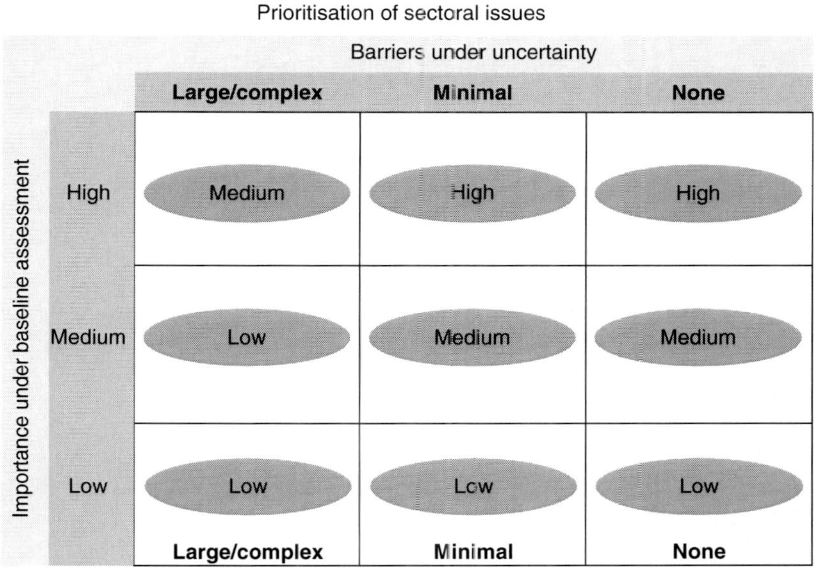

Prioritisation of sectoral issues

Barriers under uncertainty

Figure 9.2 Prioritisation matrix
Source: Developed by the authors during the SAPCC Process using multiple criteria.

will help it to tap additional resources for mitigation either from project-based mechanisms or through central funding from some of the national missions. The actions are prioritised based on different criteria, whether state-wide or area-specific, type of barriers and nature of interventions (Table 9.3).

However, this budget is an estimate and includes the 'business as usual' (regular programme budget) as well as additional fund requirements for funding adaptation. The state also relies heavily on mechanisms such as transfer of financial resources available under national missions, market mechanisms like the Clean Development Mechanism under the Kyoto Protocol, reducing emissions from reduced emission for deforestation and degradation, and green compensation to achieve its strategic objective of CC adaptation and mitigation. The state has already made provisions and factored CC in as a separate item in the annual planning and budgeting exercise, and it is probably the first state in the country to do so and presented this to the Planning Commission of the government of India.

6. Discussion and analysis

In terms of methodology, a wider consultation process is necessary to achieve an integrated CC assessment (Paul, 2001) at the state level to

Table 9.3 Summary of climate actions and budget in Meghalaya

Sectors	Total no of actions			Priority actions		Budget Million USD
	Adaptation (AD)	Mitigation (MI)	Total (AD)	Adaptation (MI)	Mitigation	
Agriculture	18	3	21	12	2	255.28
Energy	3	11	14	3	9	448.39
Forestry	16	3	19	11	3	198.28
Sustainable Habitat	10	8	18	4	4	281.20
Mining	11	5	16	7	4	18.47
Water	12	3	15	12	3	57.80
Total	70	33	103	49	25	1259.42

[Note: 1 crore = 10 million and 1 USD = 50 INR]
Source: Government of Meghalaya (2012). *State Action Plan on Climate Change [draft]*, (Shillong: Department of Planning), p.87.

reflect the state-specific issues. This discussion opens the floor up for further debate on many issues relating to CC-related vulnerabilities. The participants generate various solutions focusing on these issues. Since many of the solutions may not be easy to implement, the actions are prioritised using multicriteria-based analysis.

The time and effort of various technical experts to constructively deliberate on the various state priorities, and seek conscious alignment with the national priorities without compromising state-specific concerns, proved difficult in practice.

An attempt has been made in the formulation of the SAPCC to take into account the sectoral vulnerability, the stressors and the nature of actions. Apart from global stressors due to CC, some dynamic stressors evolve in Meghalaya due to rapid urbanisation and challenges in transport planning, and also water-sharing agreements between different states and in the region. In this example a key issue relating to sustainable habitats is resource congestion due to the high level of urbanisation. This accumulates more waste, creates pressure on transport infrastructure leading to greater emissions and pollution, and so on. Unchecked growth also makes adaptation and mitigation options politically cumbersome. The solutions have been prioritised using multiple criteria where barriers and impacts have been cross-matched based on the nature of actions, the timeframe and the scale.

The research community recognises the fact that the integrated assessment of this nature is fraught with infirmity without clearly identifiable targets. But several global, national and regional stressors have impacts on climate policies at each level (K. O'Brien *et al.*, 2004), and there is a systematic approach available in each sector to deal with them and to develop a time-bound action plan. Therefore, in this planning process, an attempt was made to begin with vulnerability mapping based on an evidence-based local-level case for each sector within the state. Subsequently the CC adaptation and mitigation actions were framed for targeting policy interventions.

Fiscal prudence, growth objectives, political compulsions and local geopolitics also act as deterrents in the planning process, though effort has been made to use the tools to choose actions that meet these criteria.

7. Limitations

The main limitation in this study is to get the requisite climate data at the subnational level to conduct a vulnerability analysis. The second important issue is to recognise the cross-sectoral linkage and conflicts, such as agriculture-forestry (diversion), energy-water (prioritisation for hydroelectric projects – a clean generation option or to provide water for agriculture in adequate quantities), and issues relating to capacity-building and finalising verifiable indicators.

There are genuine problems in engaging diverse groups of stakeholders in hammering out a solution for a very complex problem like CC mired with uncertainty. Last but not the least, the lack of scientific data and research at the local level for long-term projections acts as a major deterrent to CC action planning.

8. Conclusion

A CC action plan document, whether at the federal or state level, is a strategic document. It should not be cast in stone and should evolve continuously with emerging challenges that the state and the region may confront from time to time. All of the existing sectoral policies may need to be re-examined in the face of CC as it has impacts that cut across sectors.

Action plan implementation requires the involvement of government officials, especially the sectoral technical cadres, senior officers and politicians who are involved in setting the agenda and direction. Civil society and the private sector, which also have a significant role to play, must contribute in tandem if some tangible outcome is to be achieved within the set timeframe. In fact, during the consultation process, civil society was quite vocal about various safeguards (e.g. the protection of forests, biodiversity and traditional water bodies) while the private sector also came up with options to sustain growth without compromising the environment – for example, decongestion of transport, integrated water management and clean development mechanisms.

Climate negotiations are always controversial among countries at the global level, among neighbouring countries within a region and among states within a country. Among the mitigation actions, greenhouse gas-reduction strategies always dwelt on maximisation of the welfare of better-off people (rich states, large capital intensive plants), paying little attention to equity (where resilient actions such as using improved smokeless cooking stoves or solar devices receive minimal attention). Similarly, within mitigation pathways, both large- and small-scale interventions had their own backers with large-scale mitigation actions favoured by states that are better off – for example, a solar energy movement in Karnataka and Gujarat – but with the large-scale generation of hydropower in the northeastern region receiving less attention.

Many believe that adaptation, a natural behaviour to counter climate stressors, is more locally grounded. The vulnerability-based analysis shows the differing capacity to adapt in different societies. In this case the less developed states with limited public investment have less capacity to adapt, and they are even more vulnerable in the face of extreme weather events. This opinion came out strongly during the consultation with civil society organisations as well as technical departments. A state like Meghalaya thus has an urgent need for special attention due to its fragile ecosystem and poor population with a heavy dependence on natural resources.

There are three distinct issues to be noted: (i) local problems, especially those relating to CC, need to be addressed locally; (ii) state-specific adaptation and mitigation actions cost money and so have to be linked to resource transfer both from the central government to states and also from one department to another at the state level; and (ii) climate-resilient actions demand the convergence of resources, and departments have to work together to achieve this without working in a silo. This is the whole spirit on which the whole design of the action plan rests.

CC debate is fraught with the challenge of sustaining the aspirational rapid economic growth, which every state desires in India, while dealing with the global threat of CC. The growth aspirations of the local youth for employment and an enhanced lifestyle require actions that are climate resilient. Some of these have been attempted in the suggested priorities.

A balanced approach will be to use this CCAP as a strategic document to build perspectives, use verifiable indicators to see progress, build structures and processes, and use skilled manpower to take this forward. Whether it is mitigation actions or any adaptation pathway, the core principles should be equity, fairness and a transparent process of access to natural resources and their distribution. A mechanism/safeguard needs to be in place where such inequities can be mitigated by making additional resources available beyond the 'business as usual scenario'.

During departmental consultation, it was observed that many departments work in silos without realising that CC cuts across several departments and that they need to work together to resolve conflicts – for example, hydropower generation, agriculture and water are related but also have several conflicting areas. Several departments can attempt integrated planning for climate-resilient infrastructure, green skills, and climate-friendly products (organic and carbon-neutral products) and services.

Effort should be made to build a knowledge repository (both indigenous and modern knowledge in a usable form) at the state level or at least at the local level with multistakeholder partnerships (e.g. government, civil society, community-based organisations and the private sector). In the priority actions, especially in agriculture, forestry and the water sector, such attempts have been made to include modern and indigenous adaptation and mitigation options.

More scientific research is needed to reduce uncertainty and vulnerability, making the local environment (e.g. state, district or village) liveable and the community more resilient and better aware of the threat of CC than before, and much better prepared to cope and adapt.

The expectation in this chapter is to show that the reader knows how a CC action plan having a national mandate is debated at the state level. This process tries to protect the state interests and actions are prioritised from a resource-devolution point of view. Only if such actions are verifiable and the deployment of such resources are logical and proper can the goal of CC adaptation and mitigation be achieved in the short term, medium term and long term.

Abbreviations

CC	climate change
CCAP	Climate Change Action Planning Process
MSCC&SD	Meghalaya State Council on Climate Change and Sustainable Development
NAPCC	National Action Plan for Climate Change
SAPCC	State Action Plan on Climate Change

References

C. Boehringer, C. Fischer and K. Rosendahl (2010) 'The Global Effects of Subglobal Climate Policies', *BE Journal of Economic Analysis and Policy*, 10, 2, 13.

P. Buys, P.U. Deichmann, C. Meisner, T.T. That and D. Wheeler (2007) *Country Stakes in Climate Change Negotiations: Two Dimensions of Vulnerability*, World Bank Policy Research Working Paper 4,300 (Washington D.C.: The World Bank).

H-M. Füssel (2010) *Review and Quantitative Analysis of Indices of Climate Change Exposure, Adaptive Capacity, Sensitivity, and Impacts* (Washington D.C.: The World Bank).

P. Ghosh (2010) *Presentation on Climate Change: Indian Perspectives Post Copenhagen. By Prodipto Ghosh, The Energy & Resources Institute*, http:// www.cmsdata.iucn.org.

Government of India (2012) National Action Plan for Climate Change, http://www .india.gov.in/allimpfrms/alldocs/15651.doc, date accessed 17 October 2012.

Government of India (2002) *National Human Development Report of India, 2001* (New Delhi: Planning Commission).

Government of Meghalaya (2005) *State of the Environment Report 2005* (Shillong, Meghalaya: Department of Environment and Forests).

Government of Meghalaya (2006) *Statistical Abstract of Meghalaya: 2006* (Shillong, Meghalaya: Directorate of Economics and Statistics).

Government of Meghalaya (2008) *State Development Report Discussion Paper* (Shillong, Meghalaya: Labour Department).

Government of Meghalaya (2009) *Meghalaya State Development Report 2008–2009* (Shillong, Meghalaya: Department of Planning).

Government of Meghalaya (2010) *Office Memorandum on Working Groups for the Preparation of State Action Plan on Climate Change (SAPCC) for Meghalaya* (Shillong, Meghalaya: Department of Planning).

Government of Meghalaya (2012) *State Action Plan on Climate Change* [draft] (Shillong, Meghalaya: Department of Planning).

Government of Meghalaya (2012) *Statistical Handbook of Meghalaya (various issues)* (Shillong, Meghalaya: Directorate of Economics and Statistics).

Raymond Guiteras (2007) *The Impact of Climate Change on Indian Agriculture* (Mimeo: MIT Department of Economics).

Intergovernmental Panel on Climate Change (IPCC) (2007) *Climate Change 2007: Synthesis Report* (Spain: IPCC).

K.S.K. Kumar and J. Parikh (2001) 'Indian Agriculture and Climate Sensitivity', *Global Environmental Change*, 11, 2, 147–154.

Ministry of Environment and Forest (2010). *INCCA Initial 4/4 Assessment Report* (New Delhi: Ministry of Environment and Forest, Government of India).

National Portal of India (2011) International Cooperation, http://india.gov.in/sectors/water_resources/international_corp.php, date accessed 31 March 2011.

K. O'Brien, R. Leichenko, U. Kelkar, H. Venema, G. Aandahl, H. Tompkins, A. Javed, S. Bhadwal, S. Barg, L. Nygaard and J. West (2004) 'Mapping Vulnerability to Multiple Stressors: Climate Change and Globalisation in India', *Global Environmental Change*, 14, 4, 303–313.

Planning Commission (2008) *Government of India 'Eleventh Five Year Plan (2007–2012) Inclusive Growth and Social Sector, Volume I and II* (New Delhi: Oxford University Press).

Prime Minister's Council on Climate Change (2008) *National Action Plan on Climate Change, June 2008* (New Delhi: Government of India).

A. Paul (2001) 'Procedural Fairness in Economic and Social Choice: Evidence from a Survey of Voters', *Journal of Economic Psychology* 22, 2, 247–270.

B.K. Pradhan and J. Ghosh (2012) *The Impact of Carbon Taxes on Growth Emissions and Welfare in India: A CGE Analysis*. IEG Working Paper No. 315 (New Delhi: Institute of Economic Growth), p. 34.

A. Sanghi, R. Mendelsohn and A. Dinar (1998), 'Measuring the Impact of Climate Change on Indian Agriculture', in A. Dinar (ed.) *World Bank Technical Paper No 402*, (Washington D.C.: The World Bank).

D.S.G. Thomas and C. Twyman (2005) 'Equity and Justice in Climate Change Adaptation amongst Natural-resource-dependent Societies', *Global Environmental Change*, 15, 2, 115–124.

R.S.J. Tol (2001 'Equitable Cost-Benefit Analysis of Climate Change Policies', *Ecological Economics*, 36, 1, 71–85.

10
The Role of Government and the Private Sector in Mitigating and Adapting to Climate Change

Vinay Sharma

1. Introduction

Scientific data and analysis present a detailed account of several cycles of climate change (CC) since the inception of life on this planet (see http:// climate.nasa.gov/evidence/, http://www.ipcc-data.org/). There was a period when ice was dominant. Even if that altered the human race forever. Importantly then, humans could not do much about such changes. They were the spectators and not the participants, and they were not responsible for instigating or accelerating those changes.

Even today, humans are spectators but they are seriously guilty of having caused the changes that have led to environmental damage. The positive aspect of the situation is that they understand the damage that they have caused and will be able to exercise damage control if they can unanimously agree to do so. For example, referring to the contamination and pollution of the River Ganges in India and further to the website at http://gangapedia .iitk.ac.in, one realises that we do recognise the damage we are doing and how we may be able to control it.

The most important thing is that acknowledging the adverse effects of CC is now not enough. Our awareness levels are quite high and policies are in place. International negotiations with individual country perspectives and interests are going on and some results will definitely emerge. For example, the Food and Agriculture Organisation of the United States is assisting a number of countries in implementing projects related to CC (see http://www.fao.org/ climate change/projects/en/). However, these results must be complimented by all-encompassing efforts from all groups of stakeholders, especially from corporates, simply because the world is currently driven by industrial and corporate activities. For example, power generation is one of the main causes of rise in the levels of per capita

148

energy consumption, which is, in turn, the key to several other activities related to 'capability' development and enhancement and perceived development, which may bring wealth (Sen. 2000). Power generation accelerates CC either through carbon emissions or by interfering with the flow of rivers and instigating disasters, as recently witnessed in Japan with regard to nuclear energy. Power generation can only be useful if it is driven by some revolutionary technology without damaging effects Such a technology is yet to be developed and commercialised. Therefore the private sector can play a pivotal role in creating and directing a movement to control the damage.

2. Literature review

No one has the answer in immediate terms to this paradox. Whenever panic strikes, countries tend to overcome the situation by generating further awareness, entering negotiations and setting budgets with regard to curbing future effects. Ultimately the result is nothing but the popularisation of terms such as 'sustainability' (United Nations, 1987).

CC is not new to the private sector A research study conducted by Agrawala (2011) reported a 'high level of awareness' among all of the companies interviewed about the issue. A report (2010) by Asian Tiger Capital Partners for the International Finance Corporation on engaging the private sector in CC policies in Bangladesh emphasised the strategies required to engage the private sector in addressing the issue of CC because the private sector has 'particular competencies' related to innovative technology, flexible infrastructure, information systems and the capacity to make a 'unique contribution' to the cause. It also suggested the need for collaboration between the private and public sectors to achieve a 'climate-resilient' world. According to the report, both of these sectors have resources that are both complementary and supplementary to each other. Similar potential was seen earlier in the Stockholm Environment Institute report (Stockholm Environment Institute, 2009).

Win (2011) in *Urbanising Asia Unprepared for the Climate Change* has written about Asian cities overlooking environmental and CC issues while pursuing growth and sustainability in such a way that the crowded population at large has become a major issue:

> The carbon pollution by India is increasing and it is becoming one of topmost carbon polluters. But to deal with this issue, India has come up with a mechanism called PAT, i.e. Perform, Achieve and Trade, which is world's first market based mechanism to deal with pollution issue. As per this mechanism, the benchmark efficiency levels are fixed. The companies which use more energy than the fixed levels have an option of buying energy saving certificates from other companies which consume less

energy. PAT is expected to curtail approximately 100 million tonnes of carbon emission in India.

(Kumar, 2012)

Vezirgiannidou (2009) has aptly expressed the importance of credibility in setting up a compliance system to deal with CC. According to him, there is a need to increase the participation of both private and public sectors in dealing with CC on a global scale by working towards a treaty. It seems to be a good option but the success of such a treaty, like all others, will depend on the level of participation and commitment of the member states.

Ilan (2010), in a paper entitled 'Policy Arena Introduction to Climate, Disasters and International Development', explored mechanisms of disaster risk reduction related to CC and developmental policy, including various dimensions such as 'vulnerability', 'capacity', 'resilience', 'climate refugees' and 'climate conflict'. Based on the history of international developments, he suggested the need to address the root causes, such as vulnerability and poverty, and to formulate and implement an effective development policy.

Semenza, Ploubidis and George (2011) have stated that the impact of global change on human and natural systems is very severe and extensive. It affects those who are most vulnerable in a disproportionate manner. There are two different strategies of coping up with this impact – namely, mitigation and adaptation. Not only do industry, businesses and government have a role to play in implementing these strategies but an equally important part has also to be played by individuals in terms of behavioural change.

Behavioural changes can be brought about by properly educating people through different points of access, such as spiritual leaders educating their followers from the perspective of social conduct wherein they motivate them to take an intrinsic oath to commit themselves to the good of society.

Kee, MaHong and Muthukumara (2010) provide preliminary econometric evidence using multivariate data analysis. They suggest that there is a probability that domestic CC policies, which may have adverse effects on international trade – that is, internal issues related to CC – can also influence external events.

The holy River Ganges: An intense example of environmental mishandling coupled with the effects of climate change

There has been a strong belief that the River Ganges, revered to be like a mother in Hinduism, could absorb any form of pollutant and could purify anything. The concepts of *Aviral* and *Nirmal* – that is, unhindered continuous flow and purity – have always been exemplified through this wonderful source of water. Statistically, around 50 per cent of India's population lives in the vicinity of the Ganges and the Gangetic plains are a huge source of agricultural products, industrial products, fisheries, sand and tourism, to name a few. This single source of water can claim a very large contribution to the economy of India. Having contributed so much, how it is treated in return is devastating. Also, the measures which may control the damage are hugely expensive. The magnitude of the Ganges Basin (Gangapedia, 2012a), hydropower projects being supported by the river (Gangapedia, 2012b) and the present levels of pollution (Gangapedia, 2012c) are indicative of the fact that the unfathomable capacity of the Ganges to absorb pollution and pollutants has at last been exhausted. The cause is the people whom the river has been supporting and feeding. As a final insult, the 'Gangotri' glacier from where the Ganges emerges is receding and several scientific warnings have been issued.

The question is why this happened, why it was not stopped and, now, what can be done? The damage to the population that the river supports will be irreparable if nothing is done. At this stage, no clear solution seems to be available. However, there is at least greater awareness and the world is working towards solutions.

Rodin's paper 'The Next Great Challenge for the Developing World' (2008) explains that we can and must be smart about globalisation. According to him, the world must learn to exploit resources – both scientific and technological – in any part of the world to promote economic development as well as resilience.

Any perspective or interpretation which adds greater complexity to the thought process and raises simultaneous questions should be examined to ensure that the solutions obtained are tangible.

3. Research objectives

The objective of this chapter is to explain the gravity of CC (through examples, such as the River Ganges) and its widespread implications on a very large population with a socioeconomic perspective through a research-based understanding. It also aims to meaningfully propose the potential role and commitment of the private sector in moulding a future scenario in the context of sustaining damage control by CC.

4. Research methods

This chapter is based on reflexive methodology. The problem of reflexivity and the ways in which 'our subjectivity becomes entangled in the lives of others' (Denzin, 1997) has been of keen interest to researchers for at least 30 years (Hammersley and Atkinson, 1983; Denzin, 1989, 1995; Lather, 1991; Atkinson, 1992; Hobbs and May, 1993). To suit the context of the problem at hand, this study utilised the 'reflexive approach' applied extensively throughout the process of understanding and developing the findings and recommendations (Alvesson, 2003).

Reflexivity is an awareness of a broad range of insights of interpreting one's own interpretations as well as looking at one's own perspective from others' perspectives (Alvesson and Skoldberg, 2000). According to Mauthner (1998), interpretation of data, either primary or secondary, is a reflexive process in which meanings are construed rather than found. The interpretations carry the epistemological, ontological and theoretical assumptions of the researchers who developed them (Alvesson and Sköldberg, 2000). Reflexivity assumes, in contrast with popular practice, that researchers are parts of the research process. The bias of the researchers can thus be minimised through interpretation (Alvesson, 2003). The chapter, while analysing secondary data including the reports available on relevant organisations' websites, substantiated by my observations and experience during my involvement in a major project on CC, acknowledges the importance of being reflexive in research and maintains the same during the entire data analysis.

5. Findings and discussion

In sync with the research objectives, the following findings emerge. To begin with, the question that arises is why are the private sector and corporations important? There are several simple reasons:

(a) Governments are already dealing with CC but the outcome is uncertain.
(b) The private sector drives the economy and people tend to follow what institutionalised and globalised private institutions initiate because their lives are directly affected. This sounds simple but by nature it is paradoxical because for private organisations, driving and capitalising demand is the way and competition is the benchmark.

World leaders promote ambitious multilateral agenda for responsible business conduct (OECD, 2011)

Ministers from countries which are part of the Organisation for Economic Co-operation and Development (OECD) developing economies of the world together with the US secretary of state, Hillary Rodham Clinton, undertook the cause of promotion of 'a more responsible business conduct' at the OECD's 50th anniversary. The group of delegates agreed to new guidelines in this context for multinational companies. The updated OECD Guidelines for Multinational Enterprises are based on 'new, stronger standards for corporate behaviour' and included recommendations on human-rights exploitation, responsibility for supply-chain management and so on. The guidelines call for 'respect for human rights in all countries of operation by the enterprises and promotion of free and open internet' for all (OECD, 2011). The stress was on abiding by international standards, with due diligence, for issues like payment of wages, dealing with cases of bribery and extortion, and promotion of sustainability (OECD, 2011).

Pinkse and Kolk (2010) have suggested that the role of companies in addressing global change is very critical because of their innovation capacity. However, the companies have adopted a rather cautious approach in providing innovation-based support towards a low-carbon economy.

Adding to the above clues and arguments, Moise and Steenblik (2011) give a snapshot of the measures adopted for CC control related to non-product processes as well as production methods. Those most emphasised were related to the greenhouse gas emissions of various products. They also explained the significance of services like business, telecommunications services, construction and related engineering services to implement the process of mitigating emissions.

Therefore furthering the efforts made to date and capitalising upon the awareness levels generated, consensus-built projects like the Carbon Declaration Project (CDP) have been initiated Not only are the effects evident but also the objectives of such voluntary disclosure-based efforts are being analysed (see https://www.cdproject.net/en-US/Pages/ HomePage.aspx).

Stanny and Ely (2008) examined factors associated with the decisions of US S&P 500 firms to disclose CC-related information specifically requested by institutional investors. The results supported the extant literature on voluntary disclosure. According to these results the firms are subject to more scrutiny because of their size and previous disclosures, and foreign sales were those more likely to respond to the CDP questionnaire. Voluntary disclosure via the CDP questionnaire increased from 48 to 58 per cent of S&P 500 firms, indicating recognition of the importance of the information. However, the same paper poses several questions associated with the need for intensification of such efforts, investor involvement and further transparency.

> The choice here, though, has to be made in relation to enforcement, incentives, motivation, unification, universalisation through change of measurement criterion and measurement scales and parameters, with the context that the methodology should yield swiftly.

Any of the methods may work but it needs to be acknowledged that the methods adopted now will set a precedent for the future. So they should have the property of self-replication, wide acceptability and universal benefits for the betterment of our home, planet earth.

6. Recommendations

Hence this chapter proposes a three-pronged strategy:

- An overall and intensified focus on innovations and bringing out yield from such innovations, which are related to commercialisation with a mass-market approach of technologies and procedures associated with developing alternatives putting a stop to the damage done and retrieval from this stage.
- A change of measurement criterion associated with cost and profitability, wherein accounting procedures may be evolved in a manner that accommodates not only the requirements associated with the responsibility of organisations towards climate but also to yield credit for them if they do so, which may be in terms of tangible acknowledgement from the stakeholders.
- A mechanism to develop a culture of self-restraint by redefining prosperity. After all, the countries have started using measures like the Human Development Index, capability, the Happiness Index and so on. Therefore, prosperity in terms of restrained growth targeted towards sustainability can be defined and acknowledged.

7. Practical implementation

Galbraith (2010) conducted a study to explore the performance of 98 firms in 10 different countries addressing CC through specific governance practices. The findings reported better performance of non-US firms compared with US firms on governance dimensions. Further, variables such as the number of directors and an independent board chair were linked with firms that were performing better.

The CDP has instigated large organisations for voluntary disclosures. These organisations have also started bringing forth the innovative practices they have been utilising and implementing, and have been projecting those practices as value propositions. Though such organisations are still few in number, the most important part is that their brand strength and brand influence are strong. On the other hand, these organisations have huge supply-chain-management systems associated with their operations. Hence they would comfortably be in a situation to influence their suppliers and associated producers. For example, some corporations are working on converting all of their buildings, premises (e.g. Bahrain Trade Tower, ITC Hotels in India, Infosys Premises in India) and production facilities into environmentally responsible setups while utilising innovative production methods of energy consumption, conservation, reproduction, water recycling and resource recycling (as done by Patanjali Food Park in India and several others; see http://pfhpl.com/index-eng.html).

Further, adding up to these efforts, several new businesses and ventures are focusing on the environment from the very first stage. Also, many of their new efforts are focusing solely on renewable energy production. Such efforts seem promising – for example, those mentioned by 'Earth the Sequel' (see http://earththesequel.edf.org/).

As discussed earlier in this chapter, the level of contribution being made by the Ganges and the magnitude of the dependence of the population in India on this holy river calls for several measures. These are expressed in the preceding paragraphs so as to reduce the pressure of ill effects being generated until now.

Three very important and fundamental findings emerged from the research study in this regard, apart from the norms related to industrial pollution, waste discharge, sand mining, stone mining, fishing and a few other things directly affecting the Ganges. They are:

- Holiness associated with the river: the primary findings associated with the interviews conducted with the people linked to the Ganges revealed that the holy relationship between people and the river is diminishing. Hence people at large now have become insensitive towards polluting it.
- Fundamental definition of tourism: according to this, the visitors who do not stay overnight are not counted through database methods as tourists. Hence an exact picture of the number and types of people interacting with the Ganges at particular locations is not available.
- Concentrated festivals: KumbhMela, KartikPoornima, KanwarYatra and several other festivals bring millions of people to holy places in India, such as the cities of Haridwar and Allahabad, for a short period. This cannot be addressed through any methodology except for either spreading the people or spreading the visiting days.

Several policy interventions are required and deliberations are ongoing. For example, Kachel (2010) suggested the need for research which 'achieves a holistic understanding of the nature and influences of environmental learning on tourists' environmental values and travel experiences in relation to climate change'. Most quantitative studies in ecotourism focus on these values and behaviour only, but they have neglected the impact over time and tourists' learning experiences.

Subsequent Verses (Chaupaies in Hindi language have been taken from Ramcharitrmanas an Indian Epic and interpreted in English language hence forth)

श्रीरामराज्य

श्रीरामराजबैठें त्रैलोका।हरशितभए गए सबसोका।।
बयरू न करकाहुसनकोई।राम्प्रतापबिशमता खोई।।

बरनाश्रमनिजजनिज धरम।निरतबेदपथलोग।।
चलहिंसदापावहिंसुखहि।नहिंभय सोक न रोग।।
दैहिकदैविकभौतिकतापा।रामराजनहिंकाहुहिव्यापा।।
सबनरकरहिंपरस्परप्रीती।चलहिं स्वधर्मनिरतश्रुतिनीती।।
चारिउचरन धर्मजगमाहीं।पूरिरहासपनेहुँ अघनाहीं।।
रामभगतिरतनर अरू नारी।सकलपरमगति के अधिकारी।।
अल्पमष्त्युनहिंकवनिउपीरा।सबसुन्दरसबबिरूजसरीरा।।
नहिंदरिद्रकोउदुखी न दीना।नहिंकोउअबुध न लच्छनहीना।।
सबनिर्दंभ धर्मरतपुनी।नर अरू नारिचतुरसबगुनी।।
सबगुनग्य पण्डितसबग्यानी।सबकष्तग्य नहिंकपटसयानी।।
रामराजनभगेससुनुसचराचरजगमाहिं।
कालकर्मसुभावगुनकष्तदुख काहुहिनाहिं।।
बाजाररूचिर न बनइ बरनतवस्तुबिनुगथपाइए।
जहँ भूपरमानिवासतहँ की सम्पदाकिमिगाइए।।
बैठेबजाजसराफबनिकअनेकमनहुँ कुबेरते।
सबसुखीसबसच्चरितसुन्दरनारिनरसिससुजरठजे।।

Shri Ram Rajya

Every Citizen of Shri Ram's Rajya is happy, content and
crystal clear about his duties, responsibilities and direction.
Every person is aware of the range of benefits they have to
look for and expect. The people look towards the society and
the world through a very large and everlasting perspective
and on an immensely broad canvas. There is trade and
business and every other activity related to any social
structure but the system functions in a
self-propelled self-sustainable manner

Redefining prosperity and intergenerational prosperity to achieve sustenance

Another important part of the strategy proposes that the type of wealth and
prosperity which has been enjoyed by preceding generations should hold
the same meaning now as well but the meaning should now incorporate the
concept of alongside sustenance deriving the desired and augmented ends.

8. Conclusion

The human race has injured and demeaned the environment on a large
scale. Consequently, the remedy should be as elaborate as the damage.
This chapter stresses that awareness regarding a grave issue like CC is
not enough. What is needed is the formulation of effective, actionable

strategies to control the damage. While acknowledging the role of government in the process, the chapter emphasises the role of the private sector in establishing a precedent for sustainability. Redefining the inclusion of activities related to corporate social responsibility and voluntary disclosures, following pollution-related norms, renaming terms by prefixing 'green' and so on have played the role of sensitising one and all. The world is aware of what is coming up. However, some real readjustments in an integrated fashion must be made. The cooperation of government and the private sector would help to achieve the goal of developing prosperity through innovation. For example, the CDP is an intense effort to generate voluntary disclosure in an organised manner from frontrunner organisations which are known for wealth creation and market development. These organisations have been looked upon as models in terms of innovation, operational optimisation, rationalisation and efficiency. Then the ultimate goal of mutual happiness would be attained by the evolution of an augmented measurement criterion and the development of a culture driven by human-centred interrelated prosperity, and there would be a nation.

> Where the mind is without fear and the head is held high;
> Where knowledge is free;
> Where words come out from the depth of truth;
> Where tireless striving stretches its arms towards perfection;
> Where the clear stream of reason has not lost its way into the
> dreary desert sand of dead habit;
> Where the mind is led forward by thee into ever-widening thought
> and action...
>
> (Rabindra Nath Tagore, 1910)

9. Limitations and future directions of research

The limitations of this chapter include the following:

- The propositions presented are associated with the understanding that I have developed over time through my involvement in a project of national relevance in India. Therefore, to further theorise, structured primary research may help.
- A validity check of propositions made through a descriptive research-design-based study wherein the fundamentals discussed may further be broken into constructs and then analysed in different environments would enhance the understanding of the researchers and would help in universalisation.
- Analysis of the propositions with the perspective of geographies, nations and societies would further strengthen the applicability of the context presented.

Further research is associated with how the verse from ShriRamcharitrmanas can not only be validated with the understanding that people have with reference to the similar context but also how societies may experience contentment, sense of duty, responsibility and direction in a self-propelled and self-replicable manner.

Abbreviations

CC   climate change
CDP   Carbon Declaration Project

References

S. Agrawala (2011) *Private Sector Engagement in Adaptation to Climate Change: Approaches to Managing Climate Risks*, OECD Environment Working Papers (No. 39) (Paris: OECD Publishing).

M. Alvesson (2003) 'Beyond Neopositivists, Romantics, and Localists: A Reflexive Approach to Interviews in Organisational Research', *Academy of Management Review*, 28, 13–33.

M. Alvesson and K. Sköldberg (2000) *Reflexive Methodology: New Vistas for Qualitative Research* (London: Sage).

Asian Tiger Capital Partners (2010). Report on A Strategy to Engage the Private Sector in Climate Change Adaptation in Bangladesh http://www.climateinvestmentfunds .org/cif/sites/climateinvestmentfunds.org/files/IFC_pres_CC_PS_V8_Sep12010 -_IFC_%20sk.pdf, date accessed 2 October 2012.

P. Atkinson (1992) 'The Ethnography of a Medical Setting: Reading, Writing and Rhetoric', *Qualitative Health Research*, 2, 451–474.

N.K. Denzin (1989) *Interpretive Biography* (Thousand Oaks, CA: Sage).

N.K. Denzin (1997) *Interpretive Ethnography: Ethnographic Practices for the 21st Century* (London: Sage).

N.K. Denzin (1995) *The Poststructuralist Crisis in the Social Sciences* (Urbana: University of Illinois Press).

J. Galbraith (2010) 'Corporate Governance Practices that Address Climate Change: An Exploratory Study', *Business Strategy and the Environment*, 19, 335–350.

Gangapedia (2012a), Ganga River Hydro Power Projects, http://gangapedia.iitk. ac.in/sites/default/files/cmapimages/Hydro%20electric%20projects%202010.jpg, date accessed 2 October 2012.

Gangapedia (2012b), Magnitude of the Ganga Basin, http://gangapedia.iitk.ac.in/ sites/default/files/cmapimages/Ganga%20divisions%20phsiographically.jpg, date accessed 2 October 2012.

Gangapedia (2012c), Present Levels of Ganga Pollution, http://gangapedia.iitk.ac .in/sites/default/files/cmapimages/reasons%20of%20Ganga%20pollution.jpg, date accessed 2 October 2012.

M. Hammersley and P. Atkinson (1983) *Ethnography: Principles in Practice* (London: Tavistock).

D. Hobbs and T. May (1993) *Interpreting the Field: Accounts of Ethnography* (Oxford: Clarendon Press).

K. Ilan (2010) 'Policy Arena Introduction to Climate, Disasters and International Development', *Journal of International Development*, 22, 208–217.

Investment News (2011) A publication of the Investment Division – Secretariat of the OECD Investment Committee, www.oecd.org/investment. date accessed 10 September 2012.

U. Kachel (2010) 'Exploring tourists' environmental learning, values and travel experiences in relation to climate change: A postmodern constructivist research agenda', *Tourism and Hospitality Research*, 10, 130–140.

H.L. Kee, H. Ma and M. Muthukumara (2010) 'The Effects of Domestic Climate Change Measures on International Ccompetitiveness', *The World Economy*, 33, 820–829.

S. Kumar (2012) Unique Scheme for Energy Efficiency, http://www.thehindubusiness line.com/opinion/article3303950.ece, date accessed 8 October 2012.

P. Lather (1991) *Getting Smart* (New York: Routledge).

N.S. Mauthner (1998) ' "It's a woman's cry for help": A relational perspective on postnatal depression', *Feminism & Psychology*, 8, 325–355.

E. Moise and R. Steenblik (2011) 'Trade-Related Measures Based on Processes and Production Methods in the Context of Climate-Change Mitigation', OECD Trade and Environment Working Papers, 2011,'04. OECD Publishing, http://dx.doi.org/ 10.1787/5kg 6xssz26jg-en, date accessed 1 September 2012.

J. Pinkse and A. Kolk (2010) 'Challenges and trade-offs in corporate innovation for climate change', *Business Strategy and the Environment*, 19, 261–272.

J. Rodin (2008) Address on The Next Great Challenge for the Developing World, Delivered to the American Association for the Advancement of Science 2008 Annual Meeting, Boston, Massachusetts on February 15, 2008, http://www .rockefellerfoundation.org/ uploads/files/'b0off3a5-5b34-4bb4-9a46-81b5bdda8c99 -021508jr_cc.pdf, date accessed 8 October 2012.

J.C. Semenza, G.B. Ploubidis and L.A. George (2011) 'Climate Change and Climate Variability: Personal Motivation for Adaptation and Mitigation', *Environmental Health*, 10, 46.

A. Sen (2000) *Development as Freedom* (Oxford: Oxford University Press).

E. Stanny and K. Ely (2008) 'Corporate environmental disclosures about the effects of climate change', *Corporate Social Responsibility and Environmental Management*, 15, 338–348.

United Nations (1987) Report of the World Commission on Environment and Development: Our Common Future, http://www.un-documents.net/wced-ocf.htm, date accessed 8 October 2012.

Stockholm Environment Institute (2009), Private Sector Finance and Climate Change Adaptation, http://www.sei-international.org/mediamanager/documents/ Publications/Climate-mitigation-adaptation/policybrief-privatesectorfinance-adaptation.pdf, date accessed 8 October 2012.

S.E. Vezirgiannidou (2009) 'The climate change regime post-Kyoto: Why compliance is important and how to achieve it', *Global Environmental Politics*, 9, 41–63.

T.L. Win (2011) Urbanising Asia Unprepared for Climate Change, http://carbon-based-ghg.blogspot.in/2011/06/urbanising-asia-unprepared-for-climate.html, date accessed 2 October 2012.

Part III

Governance Approaches to Managing Climate Change

11
Integrated Governance and Adaptation to Climate Change

Ken Coghill and Ramanie Samaratunge

1. Introduction

The contribution of human activity to climate change (CC) and its devastating potential to cause catastrophic damage to our environment is beyond dispute (Intergovernmental Panel on Climate Change (IPCC), 2007). In addition to the urgent need to act to curb further change, adaptation to CC already occurring is required. We argue that effective action to adapt to CC requires an integrated governance approach which brings together all sections of society. Building on earlier work investigating the governance of responses to a much more sudden event, the 2004 tsunami (Samaratunge, Coghill and Herath, 2008, 2012; Samaratunge and Coghill, 2010), the chapter extends the integrated governance approach to adaptation to the slower but more widespread and devastating effects of CC.

2. Context

CC is at once both global and local. It is global in that it is a consequence of changes in the one borderless atmosphere shared by every part of the globe, from the surface of the sea to above and beyond the highest peaks. Every discharge of CO_2 and other greenhouse gases is quickly mixed through the atmosphere that we all share. CC is local because every emission arises from a local source. It is local because its effects are felt locally, whether through more severe weather events – floods, droughts – or changes in rainfall patterns.

CC is not merely a policy issue; it is a product of pollution of the atmosphere. There is a strong argument that this pollution is the subject of international law. One of the leading legal authorities of our time, Weeramantry of the International Court of Justice (1997), argues:

> The protection of the environment is likewise a vital part of contemporary human rights doctrine, for it is a sine qua non for numerous

human rights such as the right to health and the right to life itself. It is scarcely necessary to elaborate on this, as damage to the environment can impair and undermine all the human rights spoken of in the Universal Declaration and other human rights instruments. (92)

This reflects the earlier United Nations (UN) General Assembly Resolution 43/53 in 1988. Paragraph 1 of the resolution states that 'climate change is a common concern of mankind, since climate is an essential condition which sustains life on earth' (UN General Assembly, 1988, p. 133). Other authorities argue that every government exercises a public trust under which it must protect the atmosphere. This public trust not only operates domestically, within national borders, but extends to an obligation to not cause harm to other nations (Wood, 2009). The argument is based on principles of sovereign trust obligation and founded on the UN Framework Convention on Climate Change (UNFCCC) – ratified and in force in most countries. These countries committed themselves to 'protect the climate system for the benefit of present and future generations of humankind' (UNFCCC, 2013).

In a similar way, leading Canadian authority Fox-Decent argues for a 'fiduciary theory of human rights', which includes 'the right to a healthy environment' as a human right (Fox-Decent, 2012, p. 254). These contemporary conceptions of a responsibility to protect a healthy environment have parallels in ancient traditions in many societies, preserved in the value systems of indigenous communities. According to these value systems, the object of life in these communities is to preserve that natural environment upon which they depend for their survival. Thus traditional Australian Aboriginal culture – probably the world's oldest continuous culture – is founded on 'caring for country'. By this it meant maintaining the landscape with its plants and animals in their current state rather than 'improving', developing or otherwise changing the natural environment (Davis, 2009).

A description by Ross (1992) of a fictional Aboriginal woman's place in the world illustrates the point. The woman regards herself as an integral part of the natural environment which nourishes her family. In no sense is she an owner of the land. Rather, in her worldview, 'culture and environment are inseparable' and she 'know[s] deep responsibilities towards' the land (Ross, 1992, p. 149).

Many other ancient, indigenous cultures enjoy a similar relationship with the natural environment. These value systems challenge the very basis of the values that have brought CC upon us. Consider the values that drive 'modern' societies:

The development ideology is central to contemporary Western culture, and the motivating force underlying the world industrial and trade economy. In its strongest form, 'development' is treated is an imperative – because a resource is there, it must be developed if the economic conditions (prices and cost structures) are suitable. In material terms, Western cultures have become dependent on a complex web of international transactions which distance them many steps from the environments from which their materials actually originate. The system meets not only people's subsistence and material needs, but has also come to underpin their demographic and social organisation.

(Ross, 1992, p. 151)

These 'modern' values are founded on a belief in a peculiar notion of progress: progress which falsely assumes that the resources on which it depends are inexhaustible and infinite (Wright, 2004). Most fundamentally, those resources include the atmosphere. Only now, some two centuries after severe pollution of the atmosphere began during the Industrial Revolution in Europe and a half-century after the evidence of CC began emerging, are we realising the harm that is occurring. This paper very deliberately says 'we' because it addresses an audience of public-policy and public-management practitioners and scholars. We are the people who must provide advice and leadership in our societies.

In providing that advice and leadership, we must be prepared to examine and challenge both our own beliefs and values, and those of the local, national and international communities that we serve. One of these is the assumption that progress in the form of ever-increasing consumption of non-renewable resources will necessarily lead to better lives for the people. In fact, there is little evidence that people are more satisfied with increased consumption, above certain basic levels of essentials such as food and shelter. Consider Table 11.1, which compares life satisfaction in a range of countries with the estimates of CO_2 emissions per capita and purchasing power parity per capita. CO_2 can be regarded as a proxy for the consumption of material goods, although it should be noted that some countries (e.g. the UK) have reduced their emissions by exporting energy- and CO_2-intensive industries. Life satisfaction is measured using the Cantril ladder in which those surveyed are asked to self-assess their satisfaction with life on a scale of 0 to 10, 0 being least satisfied.

Table 11.1 and Figure 11.1 show that there is no consistent relationship between people's life satisfaction and either CO_2 emissions or per capita income. We can draw two conclusions from this information. First, life satisfaction has only a limited relationship with per capita income. In other words, it is possible without the high incomes found in many developed countries. Second, life satisfaction is not a product of CO_2 emission levels. It is possible to have high levels of life satisfaction with relatively low emission levels. Knight and Rosa (2011) cite findings showing that

(1) increased consumption does not lead to higher levels of, and may actually decrease, well-being in high consumption countries, (2) high

Table 11.1   Life satisfaction, CO_2 emissions per capita and income (PPP) per capita –
selected countries

Country	Life satisfaction *Cantril ladder (0–10), mean value in 2010*[*]	Per capita CO_2 emissions from the consumption of energy (*metric tons of carbon dioxide per person) 2009*[**]	Income (PPP) (*international dollars, 2010*)[***]
Denmark	7.8	9.01	46,789
Canada	7.7	16.15	49,226
Norway	7.6	8.49	50,494
Australia	7.5	19.64	49,757
Netherlands	7.5	14.89	46,721
Sweden	7.5	5.58	50,426
Switzerland	7.5	6.00	41,988
Finland	7.4	9.93	51,128
Israel	7.4	9.74	43,703
Austria	7.3	8.43	48,487
Ireland	7.3	8.79	55,068
New Zealand	7.2	9.28	37,101
United States	7.2	17.67	67,319
Luxembourg	7.1	21.51	52,585
United Kingdom	7.0	8.35	47,986
Belgium	6.9	13.19	54,555
Iceland	6.9	11.12	43,847
Brazil	6.8	2.11	13,500
France	6.8	6.30	52,331
Mexico	6.8	3.99	19,892
Germany	6.7	9.30	42,560
Chile	6.6	3.96	32,102
Italy	6.4	7.01	45,047
Czech Republic	6.2	9.33	26,433
Spain	6.2	7.13	41,679
Japan	6.1	8.64	44,702
Korea, South	6.1	10.89	44,251
Slovak Republic	6.1	6.54	32,767
Slovenia	6.1	8.66	36,478
Greece	5.8	9.35	33,294
Poland	5.8	7.43	26,087
Indonesia	5.5	1.72	10,474
Turkey	5.5	3.29	27,671
Russia	5.3	11.23	18,268
South Africa	5.2	9.18	13,150
Estonia	5.1	13.46	44,605
India	5.0	1.38	8,496
Portugal	4.9	5.28	31,110
China	4.7	5.83	13,045
Hungary	4.7	5.00	20,831

Sources: *OECD (2011); **US Energy Information Administration (2012); ***World Bank (2012).

consumption levels are not a necessary condition for high levels of well-being, and (3) very large increases in consumption since 1961 have not led to substantial improvements in life satisfaction in the US.

Rather, the Gallup World Poll of over 140,000 people found that 'the happiest nations are those with strong social support from family and friends, freedom in making life choices, and low levels of corruption' (Helliwell *et al.*, 2010, p. 1).

Accordingly, communities which live in harmony with the environment, drawing only sustainable resources from it, may be better able to handle CC than those dependent on the extraction and exploitation of non-renewable resources (e.g. oil-based products) or harvesting renewable resources (e.g. fisheries and forests) at a faster rate than they can be renewed.

This suggests that the North with its unsustainable consumption of non-renewable resources may have much to learn from examples of low-carbon economies in the South, which nonetheless provide high levels of life satisfaction.

Some moves to recognise sustainability as a national constitutional principle have been made. Germany has included Article 20a in its Basic Law, stating:

> Mindful also of its responsibility toward future generations, the state shall protect the natural bases of life by legislation and, in accordance with law and justice, by executive and judicial action, all within the framework of the constitutional order.
>
> (Federal Republic of Germany, 1993)

Ecuador and later Bolivia have adopted constitutional laws which make extensive provision seeking to protect the natural environment from harm. (Constitucion Del Ecuador, 2008; Buxton, 2011).

3. Conservatism vs. creativity

However, a note of caution is necessary. Communities whose cultural values are founded on making no change implicitly rely upon their environment being stable. These cultures are inherently highly conservative. Conservative cultures may inhibit the creativity and innovation necessary for adaptive change to mitigate the effects of or to curb CC. In the event of the environment changing, for example, as a consequence of CC, they may be ill-equipped to adapt to changes in their environment.

4. Creativity and innovation

Societies rely on new ideas for the policies and management required for adaptation. These ideas arise from creative processes and become

168

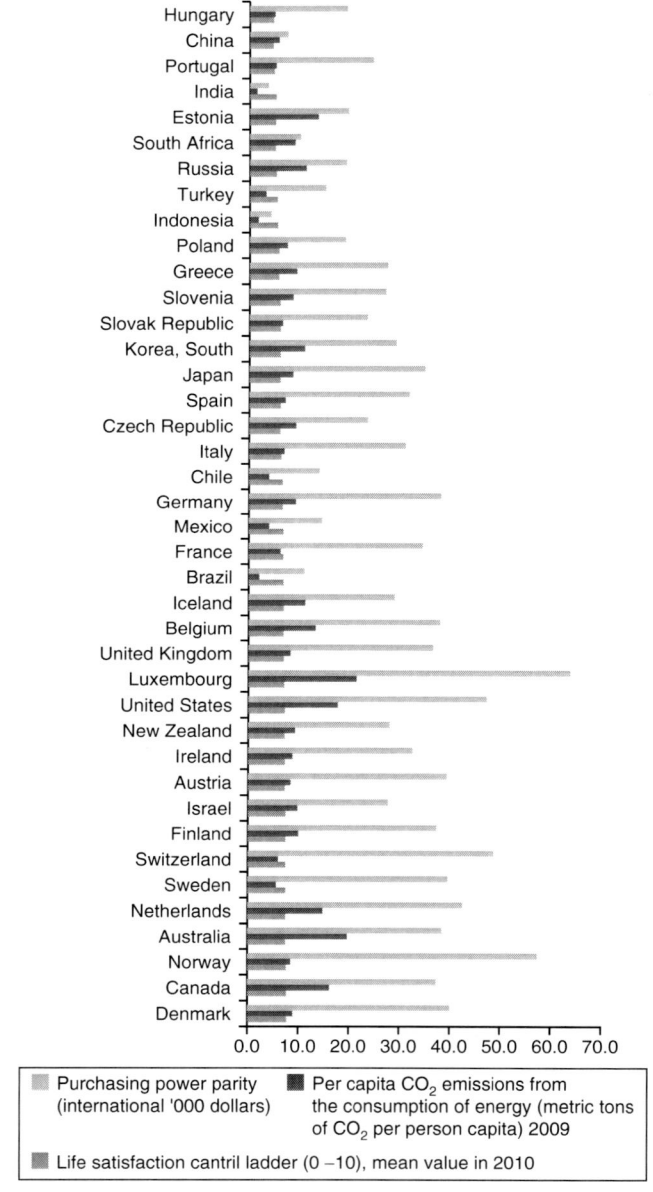

Figure 11.1 Comparing life satisfaction, CO_2 emissions per capita and income (PPP) per capita – selected countries

Sources: *OECD (2011); **US Energy Information Administration (2012); ***World Bank (2012).

incorporated in innovations. The aim of these innovations is to enable society to adapt to some circumstances, such as CC due to global warming.

Generating creative ideas, incorporating them into innovative policy, practice or management, and applying innovation to adapt to opportunities and changes in the policy environment each involve different, specific institutional culture, structure and management. Creativity ultimately arises from the thoughts of individual people. Cave (1999) believes that individual creativity 'takes place unavoidably inside our own personal, social and cultural boundaries' (p. 2). He holds that creative output requires both divergent and convergent thinking. Divergent thinking 'is the intellectual ability to think of many original, diverse and elaborate ideas', while convergent reasoning involves the 'ability to logically evaluate, critique and choose the best idea from a selection of ideas' (Cave, 1999, p. 2). Divergent thinking may include the reassembly of pre-existing knowledge, theoretical concepts or ideas in new ways, the discovery of new knowledge (e.g. through a scientific research) or the combination of new knowledge with pre-existing ideas.

Thinking beyond the individual person's creative processes, it is seen that within groups of individuals, the interactions between them sparks both divergent and convergent thinking. The essence of intellectual discourse includes the propagation and elaboration of new ideas and theories and their rigorous analysis, leading to new insights. Similarly, the same processes are seen in management and politics, although not necessarily being subject to the same intellectual rigour.

Creativity is thus an emergent process through which society's ideas arise from spontaneity, interaction and reflection. The application of creative ideas in turn enables the emergence of innovative responses to changes in society's environment (Coghill, 2004). Hirst, Kippenberg and Zhou (2009) have demonstrated that while some individuals are particularly creative, the creativity of others can be inhibited by a team-learning environment that is not conducive to their creative thinking.

There are some paradoxes revealed by the work of Hirst *et al.* (2009). A team may be more creative when it is more homogenous, particularly in the sense of functioning in more of a 'closed shop'. However, the risk of a closed shop is obvious: it is more isolated from external factors that may be relevant to the organisation and it is inherently less exposed to external sources of creative stimulus. Thus a team may be very creative in addressing issues within its field of interest but unresponsive to more significant issues that are not 'on its radar'.

4.1 Creativity leading to innovation

Innovation 'is defined as new to the unit of adoption' (West, 2002, p. 416) – that is, in the case of a society, a policy or practice that is new to the society. West emphasises the importance of innovation as the 'implementation of ideas into practice' (West, 2002, p. 411).

Innovation is therefore quite different and much more focused than the creative process. As such it calls for a culture and institutional arrangements orientated to taking on new ideas and incorporating them into reforms. There are fewer requirements for freedom to think laterally and greater need for personnel and their employees to concentrate and collaborate on the implementation process.

4.2   Adaptation

In our consideration of governance, the objective of innovation is policy, practice and management that will improve the outcomes for the society and, by extension, the global community – that is, it will assist society's adaptation to opportunities and challenges.

The success of innovation is not guaranteed. There can be no certainty that the creative process has generated effective proposals or that the innovative process has been designed so as to effectively address the policy issue. Accordingly, the final stage of the process is feedback through evaluation. The characteristics of more sustainable communities extend to their governance.

5. Governance for innovation

Field and research studies have applied integrated governance as a conceptual model to improve understanding of factors facilitating enhanced responses to severe events, such as the tsunami which struck parts of the Sri Lankan coast in 2004, killing over 30,000 people and wreaking havoc on the lives of survivors and their communities (Samaratunge *et al.*, 2008, 2012; Samaratunge and Coghill, 2010).

In this model, societies comprise three functional sectors, as represented in Figure 11.2, each of which may play roles in governance. The sectors vary in relative size between countries and over time – for example, the market sector is relatively larger in Australia than Sri Lanka; and the Australian state sector was larger relative to the market sector in the past than it is now.

These social sectors interact with each other, as shown in Figure 11.2. This model is a complex evolving system (CES) (Mitleton-Kelly, 2003). Mitleton-Kelly has described the characteristics of CESs, which we now outline.

5.1   Self-organisation

The capacity for self-organisation within the limits imposed by the system within which they are nested is a fundamental property of CESs. Societies' capacity for self-organisation is generally defined by constitutional provisions. These rules which affect the operation of a CES are of fundamental importance to its capacity for self-organisation and its outcomes. The manner in which the behaviour of a CES changes is governed by these formal rules as well as conventions and other informal controls. CESs operate much

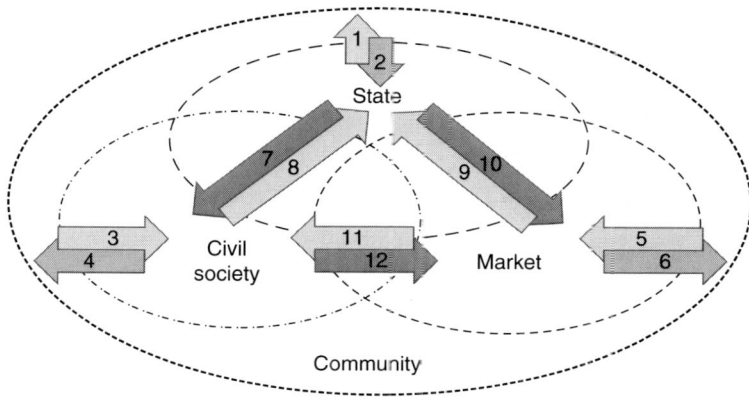

Figure 11.2 Interactions between the three functional sectors of society

more successfully where there is a moderate level of control – an intermediate point between total order and total disorder – between total centralised control and an absence of any centralised control. High levels of control inhibit the emergence of responses and suppress adaptation to change at the 'transition between order and chaos . . . excellent solutions are found rapidly'. Conversely, in 'a chaotic regime . . . no solution is ever agreed on' (Kauffman, 1995, p. 247).

Where there is very little self-regulation of the system as a whole, the law of increasing returns (Arthur, 1994) asserts itself. In other words, those with relatively more power within society tend to increase their own power at the expense of those with less power. Power tends to aggregate and accumulate. We observe a tendency of those exercising power to use it in their own personal, peer-group or class interests rather than necessarily for the overall benefit of the polity. The historic trend towards the regulation and democratisation of power in which there is a long-term trend of diminution of reliance by the state on coercive power is a response to this.

Conversely, where leaders seek to impose extensive regulatory control over society, that introduces rigidity into society, which has the effect of suppressing creativity and innovative adaptation to changes affecting its internal or external environment.

This becomes important in the design of democratic systems. Where the constitutional structure makes it unlikely that a leader or political party has unchallenged power over executive actions, then it is less likely that the executive will act in the interests of particular classes of people rather than exercise a public trust and recognise a fiduciary duty to the public in general. Where the constitutional structure or electoral system enables the executive to act with disregard to the public interest, there is a risk that some leaders will exploit the opportunity if not constrained by strong normative features of the polity. The law of increasing returns suggests that an executive will seek majority support in parliament, with the effect, if not the overt intent, of diminishing its accountability to parliament. In the same way, the manner in which a state's self-organisation is implemented can affect or limit the capacity of communities within that state to self-organise. Local government is generally subordinate to the national or subnational government and operates within the provisions of the local government legislation.

5.2 Emergence

Emergence refers to the potential for new ideas or properties to develop spontaneously from within the system or through interaction with actors (individual or institutional) outside the system. As noted above, emergence is affected by the level of control governing the system and its interactions. Teisman and Edelenbo have shown that the nature and extent of interactions between actors, including informal interactions, is a key factor affecting the emergence of solutions to policy problems. Where there is a high level of interaction between agencies (through their personnel), solutions are more likely to emerge. Where interactions are more restricted or limited to formal exchanges, better solutions are less likely to emerge (Teisman and Edelenbo, 2011).

Accordingly, the emergence of creative ideas and innovative solutions to unanticipated issues that will arise in the years, decades and centuries to come should be facilitated. Teisman and Edelenbo's findings confirm that these features are as much normative as structural. In the same way, interaction within the polity and indeed with neighbouring national and international polities should be enabled.

5.3 Connectivity

While it is trite to note that everything is connected to everything else, it is important to be aware of the significance and consequences of connections between actors. These connections may be as basic as shared language or extend to shared belief systems. 'Connectivity may also be formal or informal, designed or un-designed, implicit with tacit connections or explicit' (Mitleton-Kelly, 2003, p. 6). Take something as basic as the very purpose of governance. Many may say that its purpose is progress – the improvement of social and economic conditions. However, Davis (2009) reports:

> In the Aboriginal universe ... There is no notion of linear progression, no goal of improvement, no idealisation of the possibility of change. To the contrary, the entire logos of the Dreaming is stasis, constancy, balance, and consistency. The entire purpose of humanity is not to improve anything. It is to engage in the ritual and ceremonial activities deemed to be essential for the maintenance of the world precisely as it was at the moment of creation. (p. 158)

It is immediately apparent that someone holding that fundamental belief is going to have difficulty connecting with the miner who thinks he has some right to relocate the minerals that constitute that part of the Aboriginal world. Nonetheless, both are capable of modifying their beliefs and how they interpret and express them. As Ostrom observed, awareness of connections linking actors facilitates learning the norms that affect the relationship. This connectivity through understanding 'the other' potentially leads to appreciating the beneficial outcomes of value to others (Ostrom, 2005).

Here it is the normative 'constitution' which is the most important aspect of a constitution. It must provide leadership, forums and other means to build understanding between people, communities and businesses within the state, and with those who deal with the state.

5.4 Interdependence

The actors in a CES are each to a greater or lesser extent dependent on each other. For some the relationship is remote and weak; for others it is asymmetric, as between a disempowered woman in a remote community and the executive government responsible for public services on which she relies. The asymmetry is reversed when she exercises her vote. As Mitleton-Kelly (2003) explains,

> the greater the interdependence between related systems or entities the wider the 'ripples' of perturbation or disturbance of a move or action by any one entity on all the other related entities. Such high degree of dependence may not always have beneficial effects throughout the ecosystem. When one entity tries to improve its fitness or position, this may result in a worsening condition for others. Each 'improvement' in one entity therefore may impose associated 'costs' on other entities, either within the same system or on other related systems. (p. 5)

Positive interdependence should be strengthened by providing for the rights of all sections of the community including local governments, to be entrenched, secure and treated with respect and dignity. While this is normative it is only partly so.

5.5 Feedback

In systems thinking, feedback has a meaning similar to its colloquial use but is used in a special way. As Mitleton-Kelly expresses it, 'positive (reinforcing) feedback drives change, and negative (balancing, moderating, or dampening) feedback maintains stability in a system' (Mitleton-Kelly, 2003, p. 16). We can observe it in political discourse. Negative feedback is likely to discourage a political initiative whereas positive feedback is likely to encourage the political action which it endorses.

Research tells us that people are more satisfied with their lives and with decisions affecting them where they have had opportunities to influence those decisions – not necessarily personally but via they or their peers having had the chance to do so (Arvai, 2003; Frey and Stutzer, 2000).

The corollary is that responsibility for fostering participation and rights of participation should be entrenched and participation accepted as the default normative practice.

5.6 Far from equilibrium

A governance system which is subject to change that seriously destabilises its normal operation can become far from equilibrium, in which case a relatively minor disturbance can precipitate a dramatic change to the structure and functioning of the system (Mitleton-Kelly, 2003). It may make a relatively orderly transition to a different system, as occurred after the collapse of the Soviet Union, or it may dissolve into a disorderly, unstable system from which a new order eventually emerges, as recently in Somalia.

The features of governance discussed above would increase the resilience of the governance system and its capacity to reorganise in the event of a crisis.

5.7 Space of possibilities

When a CES is confronting challenges to its established order, the range or extent of options for change open to it is referred to as the space of possibilities – that is, how much change is possible without precipitating a collapse of that order and the transition to another order. The greater the space of possibilities, the greater the capacity of the CES to evolve and adapt to changes in its environment. Accordingly, a CES will be more resilient and better able to adapt to unanticipated change if it can expand and sustain a bigger space of possibilities.

5.8 Coevolution

As one CES changes, there are flow-on effects to other CESs with which it interacts. Evolutionary changes in each neighbouring CES as it evolves and adapts to changes in its environment of themselves change the environment of each other neighbouring CES. Each CES evolves and adapts similarly to neighbouring CESs according to the levels on connectivity and interdependence between them.

The implications are largely normative. For example, there must be opportunities created for substantial interactions between a state's officials and their counterparts in other jurisdictions.

5.9 Historicity and time

Evolutionary change in a CES is affected by the historical background embedded in its structure and norms. For example, if a governance system has developed in a Westminster parliamentary system environment in which there has been no serious challenge to continuing with it and little knowledge of other models, that parliamentary model becomes well entrenched.

Any change which is contemplated takes time to implement, and circumstances which change during implementation potentially affect implementation. The length of time taken may be the time taken for cultural change to occur to enable a policy to take effect, or the time taken to train and educate a population to a certain level or in special skills.

5.10 Path dependence

Path dependence refers to features of a CES which determine how it operates and which features, despite any imperfections, are so widely accepted or deeply embedded that the transaction costs of change are believed to outweigh the expected benefits.

5.11 Creation of a new order

The creation of a new order (structure and functioning of the CES) in response to the incapacity of the preceding order to function effectively is a 'key feature' of CESs (Mitleton-Kelly, 2003, 20). As discussed above in relation to far from equilibrium, this is the capacity of a CES to undergo radical change and reform to emerge with a different structure and/or norms governing its functioning and outcomes in response to significant change in its environment. The features discussed above would increase the resilience of the governance system and its capacity to create a new order in the event of crisis.

These characteristics of a CES affect the relationships represented in Figure 11.1 by arrows. Each arrow indicates a relationship, with the arrowhead indicating the direction of power or influence of the originating sector over the target sector. Customarily, sovereign states have monopoly legislative power and exclusive power over the legitimate use of coercive force (arrows 1, 7 and 10).

However, other actors may have roles in setting rules that are followed by some actors. For example, (civil society) occupational associations establish professional standards which are accepted by practitioners and may be enforced by the relevant associations. Typically, these standards are developed through a consultative process, giving the community, and both the market sector and the state sector, opportunities to influence their final form (arrows 3, 4, 7, 8, 11 and 12).

The desirability of common technical features in a number of products has led market sector actors to agree to common technical standards which are shared by manufacturers within industries (e.g. Wi-Fi networks). Such agreed standards are voluntary but it is clearly in the interests of manufacturers to abide by their terms. In many states, competition law means that such market-sector agreements must conform to state-sector regulatory requirements (arrows 9 and 10).

International auditing standards were subject to rules set by civil society professional associations until crises demonstrated that voluntary standards were inadequate to protect the public interest. Legislative action followed to prevent catastrophic market failures by giving regulation of the operation of financial markets the force of law.

In democratic societies there is a free flow of information and argument with the total community and between all social sectors (arrows 1–12). In authoritarian regimes, all arrows except 1, 7 and 10 may be severely constrained or effectively non-existent.

The inter-relationships shown in Figure 11.1 illustrate a simple unitary state such as a city state. In a multilevel state (e.g. national and local government or national, subnational and local government) sectors may interact with each other 'horizontally' (e.g. local-local), 'vertically' (e.g. local-national) or both. Federal systems add greater complexity, but that complexity is in the mix of component powers and influences rather than fundamentally different system design.

At international levels a range of intergovernmental actors equate with the state. A small number, such as the UN Security Council, the World Trade Organisation and some European institutions, have the power to intervene in actions taken by states. In each case, this power has been voluntarily ceded by the sovereign states which have submitted to the authority of the inter-governmental actor. To date there is no such institution with the authority to act on CC. However, as discussed above, there is potential for international law to be invoked with this effect – for example, through the International Court of Justice.

6. Participation enhances governance

The overthrow of many undemocratic governments by mass uprisings in recent decades, most recently the regimes in Egypt, Tunisia and Libya, demonstrate the difficulties and risks in attempting to rule in defiance of the 'necessary correspondence between acts of governance and the equally-weighted felt interests of citizens with respect to those acts' (Saward, 1996, pp. 468–469). Opportunities for participation in decisions affecting people's own lives have been shown to be an important factor in the public acceptance of those decisions.

Coghill and Thornton (2008) report that Frey and Stutzer (2000) studied the people living in a range of Swiss cantons with basically similar governance structures but significant differences in the opportunities available to citizens to participate in policy decisions. Citizens of cantons with greater levels of democratic participation were more satisfied than their counterparts enjoying less participation. Non-citizens, who had no political rights to democratic participation, were also studied. Swiss citizen residents were more satisfied than their non-Swiss neighbours who lacked rights. People with greater opportunities to participate in the political life of the cantons in which they lived clearly had higher levels of life satisfaction.

Coghill and Thornton (2008) also reported another significant finding, by Arvai (2003, p. 281), who showed that people

> were more willing to accept decisions in which they had been involved even where the decision was not the one they preferred. This extends to people who had the opportunity to participate but chose not to. They had confidence in the process because they were able to relate to those who chose to participate. Their satisfaction with the process was more important than the actual outcome. He suggests that the benefits of participatory decision-making lie in the 'higher quality decisions that are the product of more widely accepted decision processes'.

These research findings add support to the integrated governance model as it provides for active participation by individual and institutional actors and inter-relationships between the social sectors. Further support has been provided by studies reported in the International Review of Administrative Science, summarised by Teisman and Edelenbos (2011). These studies suggest that [integrated] governance is likely to generate better outcomes where there is good interaction between actors at the boundaries of their areas of responsibility. Their observations were made in the special case of autonomous water authorities which were not subject to a hierarchy of authority. These examples of governance operating in social systems provide valuable insights into the operation of a range of social systems. In this case the findings point to features which lead to better outcomes in the context of light roles by the state sector. What they describe as synchronisation ('integration' in our terminology) between social systems is affected by three complementary principles: self-organisation and variation; thinking and acting between self and the larger whole; and, operating on and beyond the boundaries of subsystems (Teisman and Edelenbo, 2011). Where opportunities exist or are fostered for actors to interact and work together, governance is more likely to lead to overall outcomes that are beneficial to the society.

These observations are all the more valuable because they concern mankind's relationships with and within a broader natural system – that particular vital component of the total natural environment, the water subsystem. They are cases of mankind regulating their own behaviour (that is, 'self-regulation', at the social not the individual level) in order to achieve ecological sustainability within particular limits of geography, climate and social organisation – that is, socioecological sustainability.

Similarly, where interconnection reflects shared values and where actors each have some level of interdependence with other actors, superior outcomes are expected. It also shows that the obverse is also likely – that is, where actors do not work together constructively, outcomes are more likely to be inferior (Samaratunge and Coghill, 2010). These contexts in which decisions are made and actions taken are similar to the Action Arenas described by Ostrom in the work leading to her Nobel Prize (Ostrom, 2005).

As with Teisman and Edelenbos (2011), the late Elinor Ostrom (2005) based her findings on studies of how people and societies operate in a range of circumstances. These are not abstract theories as to how people should behave in some ideal society. Her research found that the most effective societies are polycentric. In the area of addressing CC, the integrated governance model was the basis for a policy initiative known as 'climate communities', introduced by the government of Victoria, Australia, in 2009. The initiative was modelled on LandCare – a highly successful programme of support for local communities to address the degradation of land in their localities (Landcare Australia, undated). The government stated that

> Climate Communities (would) provide local groups with expert advice, research and information. Ideally these groups will build local networks with councils, businesses, schools and service groups to identify suitable local projects. They will be supported by regional coordinators and grants through Sustainability Victoria. We'll encourage them to share ideas and experiences and help them to access funding from the Federal Government and other sources.
>
> We'll also track the community's progress on making cuts to emissions, both to recognise the significance of local actions and to encourage the community to continually expand its efforts.
>
> (Department of Sustainability and Environment
> (Victoria), 2010)

A change of government following the 2010 elections led to the policy being abandoned before any significant implementation had occurred. No similar policy has been substituted.

7. Conclusion

CC is already affecting mankind and threatening to wreak havoc on communities throughout the world. However, effective action is possible; low-carbon economies actually offer examples of functioning, more sustainable societies which may find it easier to address CC than economies that are reliant on high levels of consumption of carbon-based energy.

Nonetheless, addressing CC requires major reorientation of the beliefs and values that drive much of public policy and public management. Attempts to impose such radical change in thinking carry severe risks of strong resistance and failure. Leading effective action to adapt to CC in any country requires governance that involves the participation of actors throughout society. An integrated governance approach which brings together all sections of society has the potential to enable our societies to achieve the transitions necessary to curb further CC.

Abbreviations

CC	climate change
CES	Complex evolving system
IPCC	Intergovernmental Panel on Climate Change
PPP	Per capita and income
UN	United Nations
UNFCCC	United Nations Framework Convention on Climate Change

References

W.B. Arthur (1994) *Increasing Returns and Path Dependence in the Economy* (Ann Arbor: University of Michigan Press).

J.L. Arvai (2003) 'Using Risk Communication to Disclose the Outcome of a Participatory Decision-Making Process: Effects on the Perceived Acceptability of Risk-Policy Decisions', *Risk Analysis*, 23, 2, 281–289.

N. Buxton (2011) 'The Law of Mother Earth Behind Bolivia's Historic Bill', *Energy Bulletin, April 21*.

C. Cave (1999) *Creativity Web*. http://members.optusnet.com.au/~charles57/Creative/Basics/definitions.htm, date accessed 27 December 2012.

K. Coghill (2004) 'Federalism: Fuzzy Global Trends', *Australian Journal of Politics and History*, 50, 1, 41–56.

K. Coghill, and J. Thornton (2008) 'Climatecare A Future Direction for Partnerships for Socio-Ecological Sustainability. How Governance Structures can be Adapted to The Imperatives of Saving Our Environment'. *Unpublished Working Paper*. Monash University.

Constitucion Del Ecuador (2008) Derechos de la naturaleza (Chapter seven rights of nature), Ecuardo. http://www.rightsofmotherearth.com/ecuador-rights-nature/; http://www.ecuamundo1.com/lex-dura-lex/constitucion-del-ecuador/, date accessed 11 October 2012.

W. Davis (2009) *The Wayfinders* (Toronto: Anansi).

Department of Sustainability and Environment (Victoria) (2010) Climate Communities. www.climatecommunities.vic.gov.au, date accessed 13 September 2011.

Federal Republic of Germany (1993) 'BASIC LAW for the Federal Republic of Germany (Promulgated by the Parliamentary Council on May 23, 1949) (as Amended by the Unification Treaty of August 31, 1990 and Federal Statute of September 23, 1990)'. http://www.constitution.org/cons/germany.txt, date accessed 11 October 2012.

E. Fox-Decent (2012) 'From Fiduciary States to Joint Trusteeship of the Atmosphere: The Right to a Healthy Environment through a Fiduciary Prism', in K. Coghill,

C. Sampford and T. Smith (eds) *Fiduciary Duty and the Atmospheric Trust* (Farnham, UK: Ashgate).

B.S. Frey and A. Stutzer (2000) 'Happiness, Economy and Institutions', *The Economic Journal*, 110(October), 918–938.

G. Hirst, D.V. Kippenberg and J. Zhou (2009) 'A Cross-level Perspective on Employee Creativity: Goal Orientation, Team Learning Behaviour, and Individual Creativity', *Academy of Management Journal*, 52, 2, 280–293.

J. Helliwell, C. Barrington-Leigh, A. Harris, and H. Huang (2010) *International Evidence on The Social Context of Wellbeing*. http://www.voxeu.org/index.php?q=node/4924, date accessed 26 December 2012.

International Court of Justice. (1997) 'Separate Opinion of Vice-President Weeramantry. Gabčíkovo-Nagymaros Project (Hungary/Slovakia)'. http://www.icj-cij.org/docket/ files/92/7383.pdf, date accessed 26 March 2013.

IPCC (Intergovernmental Panel on Climate Change). (2007) '*Summary for Policymakers of the Synthesis Report of the IPCC Fourth Assessment Report*'. Draft copy November 16, 2007 (Intergovernmental Panel on Climate Change).

S. Kauffman (1995) *At Home in the Universe. The Search for Laws of Complexity* (Viking).

K.W. Knight, and E.A. Rosa (2011) 'The Environmental Efficiency of Well-Being: A Cross-National Analysis', *Social Science Research*, 40, 3, 931–949.

Landcare Australia. (undated) *Landcare Australia*. http://www.landcareonline.com/index.asp (home page), date accessed 1 June 2008.

E. Mitleton-Kelly (2003) 'Ten Principles of Complexity & Enabling Infrastructures', in E. Mitleton-Kelly (ed) *Complex Systems & Evolutionary Perspectives of Organisations: The Application of Complexity Theory to Organisations* (London: Elsevier).

OECD (2011) *OECD Launches New Report on Measuring Well-Being*. http://www.oecd .org/newsroom/oecdlaunchesnewreportonmeasuringwell-being.htm, date accessed 26 December 2012.

E. Ostrom (2005) 'Understanding the Diversity of Structured Human Interactions', in E. Ostrom (ed) *Understanding Institutional Diversity* (Princeton, NJ: Princeton University Press).

H. Ross (1992) *Culture in Ecosystems: Australian Ways*. Paper presented at the Symposium on Culture and Environment organised by UNESCO and the Indonesian National Commission for UNESCO.

R. Samaratunge and K. Coghill (2010) 'Integrated Governance in Sri Lanka: A Conceptual Myth or a Practical Reality?', in J. Vartola, I. Lumijärvi and M. Asaduzzaman (eds) *Towards Good Governance in South-Asia* (Tampere, Finland: Department of Management Studies, University of Tampere, Finland).

R. Samaratunge, K. Coghill and H.M.A. Herath (2008) 'Tsunami Engulfs Sri Lankan governance', *International Review of Administrative Sciences*, 74, 4, 677–702.

R. Samaratunge, K. Coghill and H.M.A. Herath (2012). 'Governance in Sri Lanka: Lessons from Post-tsunami Rebuilding', *South Asia: Journal of South Asian Studies*, 35, 2, 381–407.

M. Saward (1996). 'Democracy and Competing Values', *Government and Opposition*, 31, 4, 467–486.

G.R. Teisman and J. Edelenbo (2011) 'Towards a Perspective of System Synchronisation in Water Governance: a Synthesis of Empirical Lessons and Complexity Theories', *International Review of Administrative Sciences*, 77, 1, 101–118.

United Nations Framework Convention on Climate Change (UNFCCC). (2013) Framework Convention on Climate Change. http://unfccc.int/essential_background/convention/background/ items/1349.php, date accessed 26 March 2013.

United Nations General Assembly. (1988) '43/53 Protection of Global Climate for Present and Future Generations of Mankind' *General Assembly Forty-third Session. V. Resolutions adopted on the reports of the Second Committee*, 133–134. http://www.un .org/ga/search/ view_doc.asp?symbol=A/RES/43/53&Lang=E&Area=RESOLUTION, date accessed 26 March 2013.

U.S. Energy Information Administration (2012) Per Capita CO2 Emissions from the Consumption of Energy. https://docs.google.com/a/monash. edu/spreadsheet/ ccc?key=0AonYZs4MzlZbdFF1QW00ckYzOG0y WkZqcUhnNDVlSWc&hl=en#gid =1, date accessed December 26 2012.

M.A. West (2002)''Sparkling Fountains or Stagnant Ponds: An Integrative Model of Creativity and Innovation Implementation in Work Groups', *Applied Psychology: An International Review*, 51, 3, 355–387.

M.C. Wood (2009) 'Advancing the Sovereign Trust of Government to Safeguard the Environment for Present and Future Generations (Part 1) Ecological Realism and The Need for a Paradigm Shift', *Environmental Law*, 39, 1, 43–89.

World Bank (2012) GDP per Person Employed (constant 1990 PPP $). http://search .worldbank.org/data?qterm=Income%20%28PPP%29&language=EN, date accessed December 26 2012.

R. Wright (2004) *A Short History of Progress* (Melbourne: Text Publishing).

12
Climate Change Governance: The Singapore Case

Huong Ha

1. Introduction

Climate change (CC) impacts have negatively affected the economic and social welfare of millions of people. CC management and sustainable development have been pressing issues in several debates and in many research projects on contemporary political and economic conditions in the last few decades (Haque, 2005; Strandenaes, 2007).

The challenges of CC management are complicated and interdependent, and they must be addressed by several groups of stakeholders in the public and private sectors, and civil society. The resources, power and legitimacy for managing CC are dispersed among various sectors and among various groups of stakeholders within each sector. Therefore it is impossible to assume that a single group of stakeholders or a single sector can overcome all issues associated with CC and environmental degradation. In other words, governance to mitigate and adapt CC impacts is the charge of all groups of stakeholders at both national and international levels (UNDP, 1997). Successful management of CC requires effective governance measures, institutions, structure and so on, which can enhance the competencies and contributions of all groups of stakeholders. Public–private partnerships, public engagement and participation will enable stakeholders to better contribute to protect the environment. However, such cooperation should not undermine the role of government. Government is still responsible for serving its citizens. Thus an integrated governance framework, including all sectors and both regulatory and non-regulatory mechanisms, is important to address issues associated with CC, and to mitigate and adapt to CC impacts.

This chapter aims to (i) revisit the impacts of CC, (ii) introduce a governance framework for CC management, including the public and private sectors and civil society and (iii) examine the key factors affecting the success or failure of this governance framework, using Singapore as a case study.

182

Although there are many dimensions of governance, this chapter focuses on the stakeholders and their roles to minimise the adverse impact of CC. This chapter is significant because it (i) addresses the research questions from a practical perspective, and (ii) provides information for further research in governance and CC given limited studies on governance in Singapore. Finally, other city states may benefit from both positive and negative lessons drawn from the Singapore experience in terms of how to improve strengths and overcome weaknesses in terms of governance to achieve a balance between economic development and environmental protection.

2. Impacts of climate change

CC refers to the change in 'temperature, moisture, wind velocity, humidity (rainfall), cloudiness, etc.' (Cuevas, 2010, p. 32). It has increased natural vulnerability, which is considered to be one of the variables causing socio-economic and biophysical vulnerability (Fussel, 2005; Cuevas, 2010).

In terms of the biophysical aspect, CC poses a significant threat to public health and represents 'a growing contribution to the global burden of disease' (Kumaresan, Narain and Sathiakumar, 2011, p. 200). The impact of CC on health varies in different individuals, countries and regions. Usually the poor, children, females, senior citizens, the disabled and those who do not have access to healthcare would be at most risk from diseases and other health problems caused by CC (Kumaresan et al., 2011).

Regarding socioeconomic aspects, CC has affected the livelihood of millions of people and it affects business activities. It would destroy crops and reduce food productivity, which contributes to widening the income gap between the rich and the poor, and thus it contributes to increasing the number of people living under the poverty line in those affected countries (Kumaresan et al., 2011). In other words, CC has negatively impacted the sources of income, economic well-being and poverty status of the people (Cuevas, 2010). It not only affects the 'weather-related mortality...infectious diseases...and air-quality-respiratory illnesses...and crop yields and irrigation demands' (Cuevas, 2010, p. 49) but also results in adverse effects on forests, water resources, land use, coastal areas, species and natural areas. CC has huge impacts on the general population, ecosystems and businesses in many ways. It influences water resources which 'increase the vulnerability of the agricultural sector and the rural population...reducing the sustainability of rural communities and livelihoods' (Hurlbert et al., 2009, p. 119). Examples of such impacts are 'change in forest composition and shift geographic range of forests', changes in the quality and quantity of water supply, 'erosion of beaches, inundation of coastal areas, costs to defend coastal communities', and changes in ecological systems and 'loss of habitat species' (Cuevas, 2010, p. 49). The list of loss caused by CC is not exhausted. It may be longer in the context of developing countries where (i) many people do not have access to adequate healthcare and social security, (ii) policies to respond to CC are unclear, (iii) enforcement and implementation of such policies are weak and (iv) there is a lack of resources.

3. Theoretical framework

3.1   A governance model for climate change

Wiener (2001) (cited in Polonsky, Miles and Grau, 2011) commented that

> climate change is complex in many dimensions, frustrating simple and heath regulatory responses. The challenge is to design a regulatory system that matches these complex realities [with the capabilities and resources of institutions and countries], and thereby accomplishes cost-effective advances in global climate protection. (p. 370)

Many policies implemented have not addressed the global issues caused by CC, and the livelihood of millions of people and business activities are seriously affected (Polonsky *et al.*, 2011). To manage CC, a governance approach that can provide clear policies, strategies and directions regarding who is to do what and how they are going to do it is required. Accordingly, the governance of CC is discussed below.

Coghill (2003, 2004), Coghill *et al.* (2004), Kooiman (1999), Rhodes (1996) and Stocker (1998) explained that a governance system consists of interdependent and communicating networks. These networks comprise various processes, structures, mechanisms and rules which could solve sociocultural, economic and political problems. Different parts of the systems must work in a synchrony to make the system function well. Kooiman (1999) and Kooiman *et al.* (2008) explain that three main sectors are embedded in a governance system, including government, business and civil society. In other words, a governance system includes

> all those interactive arrangement in which public as well as private actors participate aimed at solving societal problems or creating societal opportunities and attending to the institutions within which these governing activities take place.
>
> (Kooiman, 1999, p. 70)

No matter from what dimension governance is examined, it is a complex system which comprises several sectors and several courses of action, and it covers all external environmental influences. The environment is dynamic and evolving, and thus governance has to respond to rapid socioeconomic and political changes caused by globalisation, technology advancement, new forms of social media, new relationships in terms of how governments interact with citizens, and how different sectors in the economy interact with each other at the local, regional and global levels (Driessen *et al.*, 2012; Fisher and Surminski, 2012; Ha, 2012; Falaschetti, 2013). These concepts and definitions are the groundwork for the development of an enhanced and integrated governance framework for CC management.

The pattern and paradigm of governance have been changing in the last few decades due to the emergence of non-state actors who have increasingly played a bigger role in public affairs. The 'new arrangements beyond the state' have paved ways for new governance concepts and new models, including not only the public sector but also the private sector and civil society organisations (CSOs) (Kolk and Pinkse, 2006, p. 419). CC is transnational and has a domino effect. Thus it requires the involvement of both private and public actors and other relevant stakeholders to tackle the global phenomenon of CC (Newell, 2008).

This chapter proposes that a governance model, including three sectors – namely, government (the state, the public sector), the private sector (business, the market) and civil society – would enhance the management of CC. Keohane and Nye (2000), and Coghill (2003, 2004) also noted that the three spheres of governance are interconnected and interactive.

Thus governance must include not only mechanisms and processes but also institutions and interaction among stakeholders through which different sectors of a governance system execute their duty and exercise their rights (UNDP, 1997). However, for the purpose of this chapter, only the role of each sector in CC governance is discussed. The three-sector governance model is shown in Figure 12.1.

3.2 Role of different sectors

3.2.1 Role of the public sector

First of all, government must set a clear legal framework for environmental protection and CC management. Given the cross-national nature of CC and other environmental issues, CC regulations would be effectively implemented through an 'internationally coordinated policy' (Polonsky *et al.*, 2011, p. 372), although it is not easy to set a policy which is accepted by respective countries and can be implemented universally and result in the same outcomes. Nonetheless, any regulatory framework, either compliance-based or market-based-cap-and-trade, must be practical and efficient in order to avoid high production costs for businesses, which are usually transferred to consumers via high prices (Polonsky *et al.*, 2011). Government can also adopt different mechanisms of policy instruments, such as 'administrative regulations' (hard and soft laws), 'technological improvement regulations'

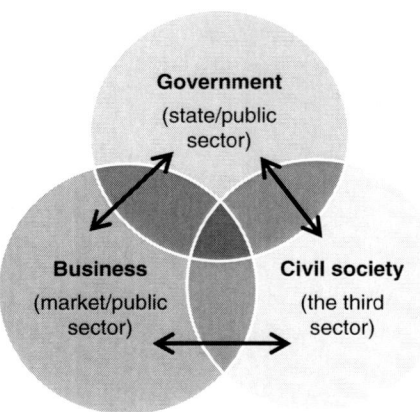

Figure 12.1 Three-sector governance model
Source: Adapted from Coghill, Ken (2004) 'Federalism: Fuzzy Global Trends', *Australian Journal of Politics and History*, 50, 1, 43.

(green technology) and 'economic instruments' (market-based or capital punishment tools, financial assistance to firms using green technology) to mitigate CC (Xu *et al.*, 2010, p. 390).

Second, government must formulate a strategic plan, incorporating long-term and comprehensive considerations regarding how to manage CC. Without a comprehensive plan, efforts to address CC may be wasted as activities may be duplicated by different groups of players (Hurlbert *et al.*, 2009).

Third, government needs to establish institutions with strong foundations and clear functions. Institutions are essential as they will manage the country's resources and build the capabilities of the public and private sectors for the management of CC (Hurlbert *et al.*, 2009). Institutions must be well established and administered in order to implement government legislation and strategies as well as coordinate the activities of different groups, in the public and private sectors and civil society. For example, the Environmental Protection Agency in the United States, and the Department of Transport, Local Government and the Region in the United Kingdom, are in charge of environmental issues in their respective countries (Christudason, 2002). In addition, sufficient funds should be invested in research and development (R&D) activities to produce innovative products and services which can reduce the amount of emissions.

Fourth, Heberle and Christensen (2011) comment that governments at the local, national, regional and global levels should be involved in the identification of the causes of CC and in finding solutions to address the root causes. Cooperation among different levels of government would facilitate the exchange of information about CC indicators and effects, and sharing experience of measures, tools, principles and standards on how to counteract the impact of CC. For instance, the government of California in the Unites States has positioned itself as 'an environmental policy innovator' and leader in establishing a governance system in which 'the state air pollution regulatory agency enjoys not only considerable control over the allocation of state funds for local air pollution control activities but also a final say in local land use planning' (Herberle and Christensen, 2011, p. 318). The relationship between government agencies at different jurisdiction levels has contributed to increasing the local communities' awareness of the consequences of CC and their willingness to tackle issues associated with environmental degradation (Herberle and Christensen, 2011).

Government must allow relevant groups of stakeholders to have a voice in the process of CC management as proposed by Hurlbert *et al.* (2009), who recommended that stakeholders should be given opportunities to be involved, provide feedback and foster learning. This can be achieved by providing education and implementing bottom-up and/or consultative approaches to engage all relevant stakeholders in pursuing a sustainable global market economy (Appleton and Lehmann, 2011), in general, and the management of CC, in particular. In Canada, relevant stakeholders, such as industry associations, non-governmental organisations (NGOs), special interest groups and water-use agencies have participated, at different levels of capacity, in the governance of water resources (Hurlbert *et al.*, 2009). Studies by Asian Tiger Capital Partners (2010) in Bangladesh, and Kreibich (2011) in Germany, also proposed that shared responsibility between the authorities and the public is required to adapt to the effects of CC.

3.2.2 Role of the private sector

Many studies have found that adopting environmentally friendly practices would help commercial organisations to deploy resources effectively and build their core competencies, which, in turn, help them to gain competitive advantages (Murad *et al.*, 2010; Jensen and Schoenberg, 2010; Wee and Quazi, 2005). There is a positive correlation between good environmental performance and the level of profitability. In other words, practising corporate social responsibility (CSR) to protect the environment does not produce any negative impact on the competitiveness of organisations (Wee and Quazi, 2005). It is important for businesses to contribute to the protection of the environment as improvement of the environment is part of the increasingly strong economic essential (Perry and Sheng, 1999). For these reasons the private sector around the world has demonstrated its efforts to tackle environmental issues since the early 1990s. One of the initiatives is the establishment of the Business Council for Sustainable Development in 1991, including members heading well-known companies, such as 'Dow, Dupont, Ciba-Geigy, Shell, Chevron, Trans Alta Utilities, Nippon Steel, ALCOA, Volkswagen and Nissan', which have provided inputs to the Rio conference (Wee and Quazi, 2005, p. 96). Apart from compliance with government regulations, self-regulatory measures have been adopted by businesses. Codes of practices and codes of conduct for better environmental behaviour for businesses have been introduced by different organisations. Some examples of voluntary guidelines and codes are the Coalition for Environmentally Responsible Economies Principles, the Business Charter for Sustainable Development, and the Global Environmental Management Initiative Principles (Wee and Quazi, 2005). The institutional members of the Investor Network on Climate Risk have collectively pressurised Fortune 500 companies to adopt good carbon-emission disclosure practices and to revise their policies towards CC and environmental sustainability as well as investment strategies (Rosen, 2010).

Private involvement has been more visible in the implementation of a consensus-building approach to reduce harmful gas emissions by multi-national corporations (MNCs) (Kulovesi, 2007). These corporations have helped to shape new practices in mitigating CC (Kolk and Pinkse, 2006). For instance, more than 50 per cent of the 218 MNCs participating in the Carbon Disclosure Project (CDP) have disclosed information about their business towards environmental protection for further research on environ-mental issues (Kolk and Pinkse, 2006). Google also launched its Clean Energy 2030 plan in 2008, a diversification from its profitable business, which aims to assist firms to 'integrate energy and environment advocacy' with their business operations (Jensen and Schoenberg, 2010, p. 126).

The private sector needs to invest in R&D in order to introduce innovative ways to minimise the amount of greenhouse gas (GHG) emissions (Kulovesi, 2007). The private sector must also observe the Kyoto Protocol framework, and comply with the national regulations and guidelines on environment protection and mitigation of CC (Kulovesi, 2007).

3.2.3   Role of civil society

Civil society has been defined as the third sector in the governance system comprising independent self-organised groups (Kaldor, 2003; Teegen, Doh and Vachani, 2004; UNDP-POGAR, 2006a), which operate relatively inde-pendently from both the public and private sectors. They have the ability and capacity to take collective action against business activities or public policies which affect the interests of their members. However, these groups do not have either an intention or the legitimacy to replace the public or private sectors (He, 2002).

Measures taken by CSOs are varied and may produce better outcomes than government regulations do. Tan and Neo (2009) commented that civil society can operate in a manner which can foster the state. These activities taken by civil society may include continuously pressurising firms to adopt green technology to reduce emissions and other causes to CC for an uncertain period of time (Asmus, Cauley and Maroney, 2006). The coverage of CSOs' activities may be subnationwide, nationwide or worldwide. The Global Reporting Initiative and the CDPs are among the initiatives of CSOs to pressure business corporations to measure and disclosure the volume of the GHG emissions released by their business activities. About 2,500 organisations in developed economies have reported their GHG emission and strategies to adapt and mitigate CC to the CDPs (Emergent Ventures International Pte Ltd, n.d.). Many CSOs, such as the US Green Building Council and the Forest Stewardship Council, have introduced and developed projects to shape markets in 'ways that further environmental goals' (Rosen, 2010, p. 23).

It is suggested that education is one of the measures to increase community awareness of the impacts of CC. Education can shift the behaviour of the general public towards environment protection and can also gain support from the communities in terms of implementation of programmes and campaigns to mitigate the negative effects of CC (Kumaresan *et al.*, 2010). Education is also a powerful tool in enabling the public to make choices of consumption in a way that leads to the reduction of the adverse effects of CC and environmental degradation (Bentley, Fien and Neil, 2004; United Nations, 2010). Thus CSOs, working with government agencies and the private sector, should encourage the public to share good practices for a sustainable 'green' lifestyle (Division of Technology, Industry and Economics [UNEP], 2011).

3.2.4 Shared responsibility

Overall, although there is no 'universally accepted perspective of the causes or impacts of climate change' (Polonsky *et al.*, 2011, p. 369), the public and the private sectors and civil society have agreed that a reduction of emissions is essential to minimise the effects of global change (Al-Amin, Jaafar and Siwar, 2010). In addition there has been an increase in international mandates by all sectors to reduce CO_2 emissions which is translated into new CC regulations (Polonsky *et al.*, 2011).

In this three-sector governance model, government has to demonstrate its obligation to citizens, and the private sector must also demonstrate its CSR in order to build consumer trust and confidence in its business activities. Further, involvement of CSOs is one of the requirements of a democratic society and for the development of good governance (UNDP-POGAR, 2006b).

4. Research method

This is a conceptual chapter. Secondary data have been collected from scholarly and non-scholarly literature, such as reports and policy papers from government and non-governmental agencies. Data have also been collected from the websites of relevant organisations at the international and national levels.

The content analysis approach has been adopted to analyse collected data regarding the role of each sector in the three-sector governance model. Factors affecting the effectiveness of the multisector governance model have also been identified via an analysis of facts and evidence from the literature review.

5. Findings and discussion

5.1 Singapore case

Singapore, a small state in Southeast Asia, has been seriously affected by CC entailing global warming due to its location – that is, 'low-lying islands' (Low and Cheong, 2008, p. 1). The country has applied various measures to address the issues of CC and environmental sustainability. The following section discusses several activities and initiatives taken by different sectors to address issues associated with CC in Singapore.

5.1.1 *Activities by the public sector*

Singapore has signed and adopted international treaties, such as the 1985 Vienna Convention for the Protection of the Ozone Layer, the 1987 Montreal Protocol on Substances that Deplete the Ozone Layer, the Stockholm Convention on Persistent Organic Pollutants, the Rotterdam Convention on the Prior Informed Consent Procedure for Certain Hazardous Chemicals and Pesticides in International Trade, the Kyoto Protocol 2006 and the Association of Southeast Asian Nations (ASEAN) Agreement on Trans boundary Haze Pollution (Ministry of the Environment and Water Resources, Singapore, 2006; Perry and Sheng, 1999; World Bank, 2009). Singapore has worked closely with (i) neighbouring countries in the Southeast Asian region, (ii) member countries of economic integration groups such as the Asia-Pacific Economic Cooperation and the ASEAN and (iii) other trading country partners to address cross-border issues related to environmental protection, global warming and CC. Singapore hosts the ASEAN Specialised Meteorological Centre, which adopts advanced technology to 'provide weather information and forecast to serve as an early warning system for imminent land or forest fires in the region' (Ministry of the Environment and Water Resources, Singapore, 2006, p. 16).

At the national level, the Singapore government has enacted both hard and soft law to manage the environment and CC. Hard law refers to environmental legislation and regulations, whereas soft law includes guidelines, action plans and recommendations. For example, Singapore enacted the Environmental Protection Control Act in 1999. It also introduced the

Singapore Green Plan in 1992. This provides directions to other stakeholders for enhancing the protection of the environment. It covers the provisions of 'environmental infrastructure, public health and environmental management' (Christudason, 2002, p. 254). This plan has been reviewed many times by four committees, and the latest Singapore Green Plan 2012, covering a range of mechanisms and strategies to enhance CC management, was introduced in 2006. The National Climate Change Strategy (in 2008) has also introduced different measures to promote the use of clean energy and minimise CO_2 emissions (Goh, 2009).

In terms of institutions, the Ministry of the Environment, established in 1972, is responsible for protecting the environment and public health in Singapore (Christudason, 2002). The National Council on the Environment, now the Singapore Environmental Council, is an independent agency that facilitates and coordinates activities of different groups in the country. Its functions are similar to the executive branch of the government – that is, to execute policies and programmes by the Ministry of the Environment (Christudason, 2002).

Singapore has introduced environmental education programmes in public schools since 1968 with the 'keep Singapore clean' campaigns (Christudason, 2002, p. 256). Many other programmes have also been launched, such as 'Tree Planting Day', 'National Recycling Day', 'Clean and Green Week' and 'Community in Bloom' (referring to community gardens) (Tan and Neo, 2009, p. 531). Yet the effectiveness of these programmes has not materialised as the general population in Singapore 'does not seem to feel involved with or affected by issues of environmental pollution' (Christudason, 2002, p. 256). Only market-based programmes, imposing monetary rewards or punishing measures, have produced positive outcomes. For instance, corrective work orders have been introduced to improve the effectiveness of the anti-litter campaign, and the emphasis on saving money has helped the public to save energy and water (Christudason, 2002). Thus a long-term strategy to educate the general public on the benefits of environmental protection should be further explored and implemented in Singapore in order to achieve the desired outcomes of CC management and environmental protection.

Singapore has invested S$500 million in R&D which aims to develop new energy technology and other solutions to adapt and mitigate CC (World Bank, 2009). The National Energy Authority has also put aside S$20 million for the Innovation for Environmental Sustainability Fund to encourage testing of new technologies, and another S$17 million was spent on the Clean Energy Research and Test-Bedding Programme in 2007 (Ministry of the Environment and Resources, n.d.). iPartners Consortium on Greenhouse Gas Reduction, a public–private initiative, was launched in 2005. Its aim is to provide solutions to clients in order to reduce CO_2 emissions in Singapore and in the region (World Bank, 2009).

5.1.2 Activities by the private sector

The private sector has demonstrated its CSR by participating in many programmes and initiatives to prevent and counter the negative effects of CC. For example, Vestas, a world-class wind-technology company, has established an R&D centre in Singapore. One of the famous car manufacturers, Rolls-Royce, has been working with many research institutes in Singapore to develop stationary fuel cells (Ministry of the Environment and Resource, n.d.). A study by Quazi in 2001 has found that several companies in Singapore participating in his survey have developed systems to manage the environment which meet the requirements of the business excellence framework introduced by the respective industry (Wee and Quazi, 2005).

Listed companies in Singapore are encouraged to adopt the voluntary guidelines issued by the Singapore Stock Exchange in 2010 to disclose information to the public about sustainability. Nonetheless, only seven listed companies have produced 18 sustainability reports to date. This figure is much smaller than the number of companies releasing their sustainability reports in China (42), in India (22 per cent of the top 100 companies) and in the United States (69) in 2010 (Emergent Ventures International Pte Ltd, n.d.). In addition, best practices in environmental management may not be adopted across the sectors and industries as there is insufficient attention given to the current and potential effects of 'environmentally conscious business practices on the manufacturers itself' (Wee and Quazi, 2005, p. 98).

5.1.3 Activities by civil society organisations

According to Lee (2000), civil groups are traditionally active in Singapore, and 'civil society was more vibrant than it is now' (Tan and Neo, 2009, p. 531). CSOs have carried out different activities to increase the public awareness of the consequences of CC. For example, the Singapore Environment Council, a non-government organisation in Singapore, has facilitated and coordinated activities to promote 'green life' among schools, communities and business. It also plays a role of match-making between various sectors and groups of stakeholders to change mindsets and attitudes towards environmental protection.

5.2 Assessment of the governance approach in Singapore

5.2.1 Factors contributing to the success of the governance approach in Singapore

Several factors affect the efficacy of the three-sector governance model in Singapore, based on the findings, include leadership commitment and competency, clear strategic plans, sufficient funds, effective enforcement of laws and public education. Efficiency and effectiveness of institutions, clear roles of relevant agencies, continuous monitoring and evaluation of the implementation of CC policies also contribute to the success of the governance of CC.

Leadership commitment is one of the requirements of any policy to be enacted and implemented successfully. A good example of a lack of leadership commitment leading to stagnancy in the development of solutions for environmental sustainability is the 'unambiguous rejection of Kyoto by the US and, until recently, Australia' (Kolk and Pinkse, 2006, p. 427). The leaders of these countries hesitated to sign the treaty, which provides an opportunity for some companies in their countries to take unfair advantages in doing business. This creates greater barriers to environmental improvement.

The public sector has to build its capabilities in order to formulate and implement environmental policies competently. They must have the ability to engage, coordinate activities of and build consensus among relevant stakeholders. The public sector must also be able to gather and mobilise information, resources and knowledge regarding environmental practices and management at the international, regional, national and local levels. Finally, the public sector must have the capacity to monitor, evaluate and modify environmental policies and programmes to meet the set objectives (Cuevas, 2010).

At both the national and corporate levels, the four key factors introduced by Makower (1994) (cited in Wee and Quazi, 2005) which can improve the effectiveness of the governance system are empowerment, education, efficiency and excellence.

Empowerment refers to the engagement of all stakeholders in the process of goal-setting, criteria and option selection, implementation and evaluation of environmental protection programmes. All sectors need to work closely with one another to share resources and expertise in order to achieve the common objectives – that is, the prevention and reduction of the negative impacts of CC. Coordination among different agencies must be effective, and information and knowledge must be shared in a timely fashion (Hurlbert *et al.*, 2009). Lack of communication among different sectors and different organisations is a barrier to successful management of CC. For instance, a lack of communication among stakeholders involved in the management of water resources is a challenge to Canada (Hurlbert *et al.*, 2009). Communication and networking with relevant stakeholders is extremely important in shaping the regional and national policies to counter the negative impact of CC. Although different sectors may have different priorities, it is imperative for them to discuss and have a common understanding of how to implement environmental strategies in a manner which can reduce the gap between the activities and outcomes of CC mitigation (Al-Amin *et al.*, 2010).

Education refers to open communication and the timely dissemination of information about environmental performance and practices to relevant internal and external stakeholders. Educational programmes to prevent pollution and reduce the waste of water and energy should be introduced widely to the public. Also, it is important for companies to provide training to employees to upgrade their skills and knowledge which can help them to fulfil their environmental responsibilities (Hurlbert *et al.*, 2009).

Efficiency encourages innovation and creativity to introduce new measures or modify current measures to improve practices of environmental management. Companies should design new products and processes which can reduce the adverse effects on the environment, such as using green technology.

Excellence can be achieved by adopting total quality management at the corporate level. Internal and external audits and benchmarking are important in the process of environmental management. They can help businesses to minimise the impact of CC and achieve international standards in environmental protection (Wee and Quazi, 2005).

In the case of Singapore, education and efficiency have been achieved, whereas the other elements require further improvement. Apart from the above determinants, which can enable or disable the multisector governance model, limited participation from the private sector and CSOs is considered to be one of the hindrances to enhance the effectiveness of this governance paradigm. Without a favourably political and socioeconomic environment and government's support, other non-state sectors may not be able to fully contribute to environmental governance in Singapore. Factors affecting the effectiveness of environmental governance in Singapore are discussed below.

5.2.2 Factors limiting the success of the governance approach in Singapore

Although many activities and programmes have been implemented by the public sector, the private sector and civil society, the success of environmental governance in Singapore has been limited due to several weaknesses.

A number of shortcomings in environmental legislation, guidelines and programmes by the Singapore government have been identified. For instance, green programmes initiated by the relevant environmental agencies do not 'incorporate ecological principles' and practices (Perry and Sheng, 1999, p. 312). Regarding government legislation, the current environmental act does not address issues associated with environmental sustainability. Also, the Singapore Green Plan and the action programmes launched earlier did not set out any strategic actions for environmental protection by some sectors, namely property management (Christudason, 2002). Nevertheless, the Singapore Green Plan 2012 does discuss measures to mitigate CC. This is reflected by the fact that Singapore has encouraged businesses to adopt good energy-management practices, and educated consumers to be more energy and fuel efficient. Singapore has also promoted the use of clean energy, such as natural gas, and innovation to explore new renewable energy sources (Ministry of the Environment and Water Resources, Singapore, 2006).

Another issue is the low level of awareness of the general public of measures to protect the environment, such as product recycling, adopting green technology and ecological preservation (Perry and Sheng, 1999). In the late

1990s, many Singaporean residents had the tendency to resist practising environmentally friendly measures, namely minimising domestic waste, saving water and energy, and buying environmentally friendly products (Perry and Sheng, 1999). Singapore has addressed this issue by launching 'reduce, reuse, recycle' campaigns which have produced favourable results – that is, the overall recycling rate in Singapore increased from 40 per cent in 2000 to 49 per cent in 2005 (Ministry of the Environment and Water Resources, Singapore, 2006). Singapore has also improved its carbon intensity by 22 per cent from 1999 to 2004, and the new target is 25 per cent in the next few years (Ministry of the Environment and Water Resources, Singapore, 2006).

Leadership commitment can be observed within the country, but there was a lack of leadership initiatives by the Singapore government in terms of working with environmental agencies in other countries (Perry and Sheng, 1999). Nevertheless, Singapore has recently become a leader in the protection of the environment. The country has offered assistance to Indonesia to contain forest fires and address the haze caused by them in Indonesia (Zengkun, 2011).

The Singapore government has implemented a top-down approach which does not emphasise public participation in the decision-making process. This is one of the barriers to environmental improvement compared with other countries in the Asia-Pacific region (Perry and Sheng, 1999). An evidence-based fact is the 'low level of environmental reporting by companies' which 'partly reflects a political environment that constraints citizen demands for information and minimises regulatory pressure on companies' (Perry and Sheng, 1999, p. 310). Nevertheless, the situation has improved now with many MNCs producing CSR reports annually. Many companies are willing to provide information to external parties for scrutiny, which demonstrates their commitment to environmental sustainability (Perry and Sheng, 1999).

In general, in the case of Singapore, a combination of hard and soft law, support by business and CSOs, and price mechanisms (via capital punishment and fees) are some of the critical success factors in CC management. Leadership commitment and sufficient resources also contribute to 'green' development in the country (Singapore Green Building Council, 2010). To sum up, the forte of Singapore in managing CC effectively includes environmental policies, strict enforcement, public trust, the ability to engage relevant stakeholders, and public and business education. Yet, it needs to overcome a number of shortcomings in order to fully achieve the outcomes of climate management and sustainable development.

6. Limitations and future research

One of the main limitations of this chapter is the lack of information about the activities of the private sector and CSOs in environmental governance. Another is that only the roles and activities of different sectors have been examined, whereas governance embraces several dimensions. Finally, the environmental governance approach has been analysed in the context of Singapore, a city state, with unique political, socioeconomic, cultural and environmental conditions.

These limitations will trigger further research on various elements within a governance system and within different contexts. Future directions of research should focus on innovative technology to produce goods and services in ways that can save scarce resources and reduce GHG emissions. How to improve cooperation among different sectors should continue to be a focus of research.

7. Conclusion

This chapter has discussed environmental governance for CC management, including the roles of the public and private sectors and civil society. CC is a global phenomenon. Thus addressing issues associated with CC is not a task of any single person, agency or country but requires shared responsibility and efforts. Also, the problems associated with CC have grown beyond the reach of the regulatory agencies. Therefore businesses are encouraged to actively contribute to reducing the adverse impacts of CC. This can be done by exercising CSR and the adopting industry codes of practice. CSOs can play the role of middlemen to bring different groups of stakeholders together. They can assist government with providing education to the public, and monitor business activities and compliance of regulations and guidelines. In short, CC management is a complicated and challenging task and must be addressed by joint efforts among sectors in the governance model.

Singapore has done well in terms of setting clear strategic plans and policies for environmental protection and CC management, strict enforcement of laws, leadership commitment and competency. However, some areas need to be improved for a better outcome of CC management. These areas are insufficient participation of stakeholders, inadequate legislation regarding CC management and so on. Singapore has been trying hard to address these issues by encouraging better public and business engagement, and cooperating with neighbouring countries and members of economic integration groups of which it is a member.

Abbreviations

ASEAN	Association of Southeast Asian Nations
CC	climate change
CDPs	Carbon Disclosure Projects
CSOs	civil society organisations
CSR	Corporate social responsibility
MNCs	Multi-national corporations
R&D	research and development

References

A.Q. Al-Amin, A.H. Jaffa and C. Sitar (2010) 'Climate Change Mitigation and Policy Concern for Prioritisation', *International Journal of Climate Change Strategies and Management*, 2, 4, 418–425.

A.E. Appleton and J.P. Lehmann (2011) 'In Pursuit of a Sustainable Global Market Economy', *Corporate Governance*, 11, 4, 327–336.

P. Asmus, H. Cauley and K. Maroney (2006) 'Turning Conflict into Cooperation', *Stanford Social Innovation Review*, 4, 3 , 52–61.

Asian Tiger Capital Partners (2010). *A Strategy to Engage the Private Sector in Climate Change Adaptation in Bangladesh* (New York: International Finance Corporation, World Bank).

M. Bentley, J. Fien and C. Neil (2004) *Sustainable Consumption: Young Australians as Agents of Change* (NSW: The National Youth Affairs, Australia).

A. Christudason (2002) 'Legislating for Environmental Practices within Residential Property Management in Singapore', *Property Management*, 20, 4, 252–263.

K. Coghill (2003) *Towards Governance for Uncertain Times: Joining up Public, Business and Civil Society Sectors*. Working Paper No 269 (Cambridge: ESRC Centre for Business Research, University of Cambridge).

K. Coghill (2004) 'Federalism: Fuzzy Global Trends', *Australian Journal of Politics and History*, 50, 1, 41–56.

K. Coghill, O.K. Tam, M. Ariff and L. Wilkins (2004) *Rating Governance in Australia*. Paper read at the *Integrated Governance Conference* (Prato, Italy: Monash Governance Research Unit, Monash University).

S.C. Cuevas (2010) 'Climate Change. Vulnerability, and Risk Linkages', *International Journal of Climate Change Strategies and Management*, 3, 1, 29–60.

Division of Technology, Industry and Economics (UNEP) (2011) *Educating for SCP towards Sustainable Lifestyles* (New York: UNEP).

P.P.J. Driessen, C. Dieperink, F. van Laerhoven, H.A.C Runhaar and W.J.V. Vermeulen (2012) 'Towards a conceptual framework for the study of shifts in modes of environmental governance – Experiences from the Netherlands', *Environmental Policy and Governance*, 22, 143–160.

Emergent Ventures International Pte Ltd (n.d.) *Business Response to Climate Change: Sustainability Reporting in Singapore* (Haryana, India: Emergent Ventures International Pte Ltd).

D. Falaschetti (2013) *Global Environmental Governance: Mechanism Design Lessons from Corporate Governance* (Bozeman MT: The Property and Environment Research Centre).

S. Fisher and S. Surminski (2012) *The Roles of Public and Private Actors in the Governance of Adaptation: The Case of Agricultural Insurance in India* (UK: Centre for Climate Change Economics and Policy, Munich Re Programme Technical, and Grantham Research Institute on Climate Change and the Environment).

H.M. Fussel (2005) *Vulnerability in Climate Change Research: A Comprehensive Conceptual Framework* (Oakland, CA: University of California).

K.C. Goh (2009) 'Snapshots from a Leading Eco-city', *Education Alliance Quarterly*, 4, 22–23.

H. Ha (2012) 'A Multi-sector Governance Model for Environmental Sustainability – Australia Case', in J.R. Barker and R. Walters (eds.) *New Zealand and Australia in Focus: Economics, the Environment and Issues in Health Care* (USA: Nova Science Publishers, Inc.), pp. 35–60.

M.S. Haque (2005) 'Governance for Sustainable Development in Southeast Asia: Means, Concerns, and Dilemmas' in A.S. Huque and H. Zafarullah (eds.) *International Development Governance* (London: Taylor & Francis), pp. 183–204.

B. He (2002) 'Civil Society and Democracy', in A. Carter and G. Stokes (eds.) *Democratic Theory and Today* (Cambridge, UK: Polity), pp. 203–227.

L.C. Heberle and I.M. Christensen (2011) 'US Environmental Governance and Local Climate Change Mitigation Policies: California's Story', *Management of Environmental Quality: An International Journal*, 22, 3, 317–329.

M. Hurlbert, H. Diaz, D.R. Corkal and J. Warren (2009) 'Climate Change and Water Governance in Saskatchewan, Canada', *International Journal of Climate Change Strategies and Management*, 1, 2, 118–132.

T. Jensen and D. Schoenberg (2010) 'Google's Clean Energy 2030 Plan: Why It Matters', in W.W. Clark (ed.) *Sustainable Communities* (New York: Springer), pp. 125–134.

M. Kaldor (2003) 'Civil Society and Accountability', *Journal of Human Development*, 4, 1, 5–27.

R.O. Keohane and J.S. Nye (2000) 'Introduction', in J.S. Nye and J.D. Donahue (eds.) *Governance in a Globalising World* (Washington, D. C.: Brookings Institution Press), pp. 1–44.

A. Kolk and J. Pinkse (2006) 'Business and Climate Change: Emergent Institutions in Global Governance', *Corporate Governance*, 8, 4, 419–429.

J. Kooiman (1999) 'Socio-political Governance', *Public Management*, 1, 1, 67–92.

J. Kooiman, M. Bavinck, R. Chuenpagdee, R. Mahon and R. Pullin (2008) 'Interactive governance and governability: An Introduction', *The Journal of Transdisciplinary Environmental Studies*, 7, 1, 1–11.

H. Kreibich (2011) 'Do Perceptions of Climate Change Influence Precautionary Measures?' *International Journal of Climate Change Strategies and Management*, 3, 2, 180–199.

K. Kulovesi (2007) 'The private sector and the implementation of the Kyoto Protocol: Experiences, challenges and prospects', *RECIEL*, 16, 2, 145–157.

J. Kumaresan, J.P. Narain and N. Sathiakumar (2011) 'Climate Change and Health in Southeast Asia', *International Journal of Climate Change Strategies and Management*, 3, 2, 200–208.

H.L. Lee (2000) 'Engaging the Citizen', in H.L. Ooi and G. Koh (eds.) *State-society relations in Singapore* (Singapore: Institute of Policy Studies), pp. 92–96.

S.C. Low and K.T. Cheong (2008) *Singapore Clean Energy Policy. Keynote Lecture at ASEAN COST+3: New Energy Forum for Sustainable Environment (NEFSE)* (Japan: Kyoto University), pp. 26–27.

Ministry of the Environment and Water Resources, Singapore (2006) *Singapore Green Plan 2012* (Singapore: Ministry of the Environment and Water Resources, Singapore).

Ministry of the Environment and Water Resources (n.d.) *Singapore's National Climate Change Strategy* (Singapore: Ministry of the Environment and Resources).

M.W. Murad, R.I. Molla, M.B. Mokhtar and M.A. Raquib (2010) 'Climate Change and Agricultural Growth: An Examination of the Link in Malaysia', *International Journal of Climate Change Strategies and Management*, 2, 4, 403–417.

P. Newell (2008) 'Civil Society, Corporate Accountability and the Politics of Climate Change', *Global Environmental Politics*, 8, 3, 122–152.

M. Perry and T.T. Sheng (1999) 'An Overview of Trends Related to Environmental Reporting on Singapore', *Environmental Management and Health*, 10, 5, 310–320.

M.J. Polonsky, M.P. Miles and S.L. Grau (2011) 'Climate Change Regulation: Implications for Business Executives', *European Business Review*, 23, 4, 358–383.

R.A.W. Rhodes (1996) 'The New Governance: Governing without Government', *Political Studies*, 44, 652–667.

C.M. Rosen (2010) 'The Role of Business Leaders in Community Sustainability Coalitions: A Historical Perspective', in W.W. Clark (ed.) *Sustainable Communities* (New York: Springer), pp. 13–28.

Singapore Green Building Council (2010) *Will it Pay to Build Green? Singapore as a Green Building Hub* (Singapore: Singapore Green Building Council).

G. Stocker (1998) *Governance as Theory: Five Propositions* (Paris: UNESCO).

J. Strandenaes (2007) *We-the Peoples, with the Environmental Sustainable Development and Democracy for a Better World* (New York: UNEP).

L.H.H. Tan and H. Neo (2009). ' "Community in Bloom": Local participation of community gardens in urban Singapore', *Local Environment*, 14, 6, 529–539.

H. Teegen, J.P. Doh and S. Vachani (2004) 'The Importance of Nongovernmental Organisations (NGOs) in Global Governance and Value Creation: An International Business research Agenda', *Journal of International Business studies*, 35, 463–483.

UNDP (1997) *Governance for Sustainable Human Development. A UNDP Policy Document: Good governance – and sustainable human development* (New York: UNDP).

UNDP-POGAR (2006a) *Democratic Governance, Participation: Civil Society* (New York: UNDP).

UNDP-POGAR (2006b) *Democratic Governance: Participation* (New York: UNDP).

United Nations (2010) *HERE and NOW! Education for Sustainable Consumption: Recommendations and Guidelines* (New York: United Nations Environmental Programme).

Y.S. Wee and H.A. Quazi (2005) 'Development and Validation of Critical Factor of Environmental Management', *Industrial Management and Data Systems*, 106, 1, 96–114.

World Bank (2009) *Climate Resilient Cities: A Primer on Reducing Vulnerabilities to Disasters* (Washington D.C.: World Bank).

B. Xu, Q. Sun, R. Wennersten and N. Brandt (2010) 'An Analysis of Chinese Policy Instruments for Climate Change Mitigation', *International Journal of Climate Change Strategies and Management*, 2, 4, 380–392.

F. Zengkun (2011) 'Singapore Has Offered to Help Indonesia Put out Fires', *The Straits Times*, September 14, 2011.

13
Governance Framework to Mitigate Climate Change: Challenges in Urbanising India

Mahendra Sethi and Subhakanta Mohapatra

1. Introduction

Global warming and climate change (CC) are a global threat to the mankind, which requires determined and collective human efforts. The international community has time and again emphasised that though the causes and impacts of CC are global in scale, the importance of local action for mitigation and adaptation is key (Organisation for Economic Co-operation and Development (OECD), 2009; United Nations (UN) Habitat, 2011). The amount of greenhouse gas (GHG) emissions of a society depends not only upon its fossil fuel use, technology and energy intensity, as widely understood, but is deeply rooted in how national and subnational governments moderate or influence personal, societal and civic decision-making. It has been observed that local bodies often lack an understanding of the international CC framework. With more than half of the world urbanised, it is how well the cities govern their jurisdictions in future that will determine the world's carbon emissions. With high population growth rates, urbanisation and economic development, India, like many other Asian and African countries, faces the challenge of interpreting the international frameworks to mitigate carbon emissions in its urban areas. Though it is well accepted that cities contribute to CC, there is limited empirical based knowledge about the contributions and appropriate models to mitigate GHG emissions (World Bank, 2010; UN Habitat, 2011). This chapter responds to this dynamic and complex situation to ascertain the contribution of urban areas in driving CC and to propose appropriate mechanisms for mitigation within the prevailing urban governance in India.

2. Research objectives

At present a structured and rational urban governance model to mitigate CC in Indian cities is virtually absent. Any futuristic mechanism has to take into account the contribution of carbon emissions and be synchronised with both the internationally accepted mitigation and the prevailing governance system in India. Our research is carried out with the following main objectives:

• to analyse how urbanisation in India is related to economic development and CC, considering their growth patterns in the last few decades and the forecast scenarios that behold the future;
• to determine the contribution of urban areas to CC; as a subset of national emissions to identify major drivers and their characteristics;
• to review the state of urban governance in India, specifically after the 74th Constitutional Amendment Act of 1993;
• to formulate suitable mechanisms to have urban governance in India that is inclusive to CC mitigation

3. Research methods

This chapter reviews the present state of urbanisation, economic development and CC in India as typical of Asian and African cities' towering challenges of high population growth rates and economic development. Time series data from Census of India, Mckinsey Global Institute, High Powered Expert Committee (HPEC) on Urban Infrastructure, Jawaharlal Nehru Urban Renewal Mission (JNNURM) Directorate under the Ministry of Urban Development (MoUD) jointly with the National Institute of Urban Affairs and Ministry of Environment and Forests (MoEF) are employed to evaluate existing and future scenarios. The chapter also reflects the current discourses in the assessment of city-wide contributions to CC, the methodologies and the inventories available.

In the absence of empirical based knowledge on the contributions, a fundamental assessment is undertaken for urban India. The research is based on theories propounded by the UN Habitat, the International Institute of Environment and Development (IIED) and the UN Population Fund (UNFPA). Utilising the national data from the Indian Network for Climate Change Assessment (INCCA) under the MoEF and for selected Indian cities accounted for by International Council for Local Environmental Initiatives (ICLEI) – South Asia, the research employs component analysis of each sector based on the location of activities that produced emissions.

As per the INCCA, of the total 1,727.7 million tons of CO_2e (CO_2 equivalent) emissions from India in 2007, 21 per cent has been estimated using the Tier I methodology of the Intergovernmental Panel on Climate Change (IPCC), 67 per cent using Tier II and 12 per cent using Tier III.

Meanwhile, the ICLEI study is based on the data collected from the engineering and administrative departments of the participating Urban Local Bodies (ULBs) to assess their energy consumption for services rendered to the citizens across the city. This study also collected data from relevant agencies responsible for energy supply to various sectors contributing to infrastructure growth within the city, such as residential, commercial, industrial and transportation – however, not owned by ULBs. The study follows the principle drawn from the World Resources Institute/the World Bank Centre for Sustainable Development/ICLEI GHG Protocol guidelines through a structured feedback process.

The second section reviews the governance system in India, particularly the urban governance model, its evolution – specifically the transformations in the last two decades – state policies and their relevance to international frameworks. The research reviews the provisions of Constitution of India, National Environmental Policy (NEP), National Action Plan for Climate Change (NAPCC), JNNURM, various government reports and other research studies to assess shortcomings and challenges within the governance to respond to mitigation framework and strategies. Considering the urbanisation dynamics and governance complex, the paper suggests suitable mechanisms to have mitigation inclusive urban governance. Like mitigation models adopted by UN Habitat and the OECD, the proposed framework follows all sectors relevant to mitigation. Concluding section derives inferences from the research study, realisation of objectives, relevance of the model, and further investigations or gaps to be addressed.

4. State of urbanisation in India

As per the UN, for the first time in history, city dwellers now outnumber rural inhabitants (UN DESA, 2011). Just 200 years ago a mere 3 per cent of the earth's inhabitants lived in cities. More developed regions are 75 per cent urbanised and growing modestly at an annual growth rate of 0.7 per cent (2005–2010), less developed regions are 45 per cent urbanised and expanding at 2.4 per cent, and the least developed countries though 29 per cent urbanised are fast multiplying at 4 per cent. Urbanisation is viewed as a serious problem that is influencing CC and is thought to be particularly affecting low- and middle-income nations. The average amount of global emissions is 5.8 tons of CO_2 per capita. It also varies across more developed, less developed and the least developed regions at 12.0, 4.3 and 0.3 tons of CO_2 per capita, respectively. Though wealth of the society and its consumption patterns influence GHG emissions, as generalisation suggests, the more urbanised the nation, the greater the amount of GHG emissions per person (Satterthwaite, 2009).

In this regard, India is urbanising at an exceptional rate. For the first time since Independence, the absolute increase in population is more in urban areas than in rural areas (Census of India, 2011). The Census of India recognises all those settlements as urban which either have a statutory status like municipal committee/corporation/notified area committee/cantonment board, estate office and so on, or that are census towns that fulfil all of the following three conditions simultaneously: (i) a population of more than 5,000, (ii) more than 75 per cent of the male working population engaged in non-agricultural activities and (iii) a density of population of more than 400 people per square kilometre.

With a total urban population of 377 million in India, urbanisation increased from 27.81 per cent in 2001 to 31.16 per cent in 2011 (MoUD, 2011). The variation may seem marginal but urban India is actually adding almost four times the Australian population every decade. The urban population is increasing at 2.76 per cent annual exponential growth rate, while the rural population is increasing at 1.15 per cent (MoUD 2011). The absolute increase in population is more in urban areas than in rural areas on account of net rural urban classification and migration (56 per cent) against natural increase (44 per cent) (MoUD, 2011).

As per a recent assessment by the Town and Country Planning Organisation (TCPO), there are 7,935 towns (4,031 statutory towns and 3,894 census towns) in India (TCPO, 2012). In 2001, the figures were 5,161 towns (3,799 statutory towns and 1,362 census towns). The statutory towns have increased by 6.37 per cent and the census towns by 185 per cent, signifying that a number of rural areas have attained urban characteristics and been designated as census towns. Out of 7,935 towns in India, 468 are Class-I (population more than 0.1 million) and 53 are million-plus cities. Almost four out of every five Class-I towns have a population of 0.1–0.5 million (TCPO, 2012). The average size of towns and cities in India has grown from 33,624 in 1961 to about 61,159 in 2011, thus clearly indicating that urbanisation in India is evident both in geographical spread and sheer volume (TCPO, 2012). It has been observed that the growth in big metros has become stagnant, while the newer and smaller ones are growing faster.

It took nearly 40 years (1971–2011) for India's urban population to rise to 270 million, but in future it may take half the time to add the same number. According to various estimates, by 2030, India's urban population will be 590 million (Mckinsey, 2010) to 600 million (MoUD, 2011) – that is, about 40 per cent of the total and break-even with the rural population by 2039.

Urban India's contribution to gross domestic product (GDP) has been rising steadily from 37.7 per cent (1970–1971) to 52 per cent (1999–2000) to 63 per cent in 2011, though the total expenditure on urban infrastructure is merely 1.59 per cent of GDP (HPEC, 2011). It is further forecasted that the urban share of GDP will be 75 per cent by 2031. Interlinkages between per capita incomes and urbanisation levels across Indian states are emerging more strongly, indicating that further economic growth will consolidate urbanisation trends.

As per the INCCA's study of GHG emissions in 2007, India ranks fifth in the world behind the United States, China, the European Union and Russia (MoEF, 2010). Interestingly, the amount of emissions from the United States and China are almost four times that of India. Another study supported by the Indian government (MoEF, 2009) predicts that emissions over the next two decades are on average bound to become threefold, from 1.72 (2007) to 5.22 billion tons of CO_2 (2030–2031), while per capita emissions will also become two-and-a-half times, from 1.5 (2007) to 5.6 tons of CO_2e (2030–2031).

Scenarios from various sources for economic development, the share of the urban population and GHG emissions are plotted in Figure 13.1. It clearly indicates that all of the issues are rapidly and dangerously converging towards an immediacy to limit CC, a situation where cities play a pivotal role in balancing economic development alongside mitigation of CC.

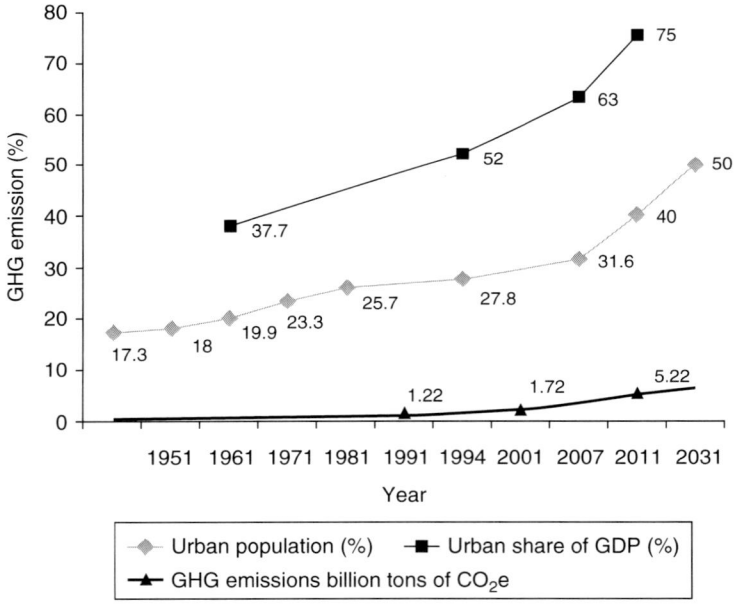

Figure 13.1 Scenarios of economic development, urbanisation and GHG emissions

Source: The time series analysis was generated by the authors based on the following data sources:
1. Urban population data from TCPO (2012), 'Data Highlights (Urban) based on Census of India', Town and Country Planning Organisation, Government of India, New Delhi. Projections for 2039 from MoUD 2011, 'India's Urban Demographic Transition'.
2. Urban share of GDP data from HPEC (2011), 'Report on Indian Urban Infrastructure and Services', The High Powered Expert Committee on Urban Infrastructure constituted by Government of India, 2008, which is based upon CSO and the 11th Five Year Plan.
3. GHG emissions data (1994 and 2007) from MoEF (2010): 'India: Greenhouse Gas Emissions 2007', Indian Network for Climate Change Assessment, MoEF, India and forecasts for 2030/2031 averaged from five models from MoEF (2009): ' India's GHG Emissions Profile', Climate Modelling Forum, MoEF, India.

Following 1991 economic liberalisation, India has witnessed phenomenal economic growth, particularly owing to the flourishing service sector in urban areas. Indian cities in the last decade have grown by 44 per cent from natural population growth, and 56 per cent from rural–urban migration and due to changes in boundary definitions (TCPO, 2012).

The analysis of 41 selected Indian cities for which data are available suggests that as the municipal area expands from 40 sq. km to 550 sq. km, the overall city emissions rise from 1 million tons of CO_2e to over 7 million tons of CO_2e (ICLEI, 2009), thereby showing a strong relationship between the level of urbanisation and fuel emissions.

5. Contribution of urban areas to climate change

There is wide discourse on the allocation of responsibility for global carbon emissions. The debate churns the international policy towards mitigation and mandates 'common but differentiated responsibilities' as protected in various protocols, summits and conventions (UNFCCC, 1992). But considering much diversity in geographical locations, economic activities, fossil fuels, technologies and consumption patterns of local populations within the same country, there is bound to be a similar argument of common but differentiated responsibilities at the subnational level. As the section on the state of urbanisation in India suggests, there is a growing shift in responsibility for CC towards city-based activities and governments.

As per the INCCA estimates, India's net GHG emissions (CO_2, methane and nitrous oxide) from anthropogenic activities in energy, industry, agriculture, waste and land use, land-use change and forestry (LULUCF) were at 1,727.71 million tons of CO_2e in 2007 (MoEF, 2010, p. 48). The methodology used by INCCA concurs with international best practices and is production based considering all emissions produced within the country's area of jurisdiction.

The emissions increased by 2.9 per cent per annum against the previous assessment of 1228.54 million tons of CO_2e (1994), while per capita emissions have gone from 1.4 tons to 1.5 tons CO_2e. Based on similar frameworks as cited in the *Cities and Climate* report (UN Habitat, 2011), contributions from urban areas have been estimated.

Table 13.1 provides the sector-wise national emissions assessed in 1994 and 2007, showing growth rates along with the estimated baseline urban contributions, with justifications of GHG contributions from the perspective of the location of the activities that produced them. Cities do not directly emit GHG emissions; it is house activities that guzzle large stocks of energy. A component-wise analysis of GHG contributions for each sector is detailed below.

Table 13.1 Contribution of Indian cities to national GHG emissions by sector

S.No.	Sector	Million tons CO$_2$e 1994	Million tons CO$_2$e 2007	Sector contribution in total Indian GHG emissions (%)	CAGR (%)	Justifications to GHG contribution from the perspective of the location of activities that produced them	Urban India baseline emissions million tons CO$_2$e 2007	Sectoral emissions at national level (%)	Sectoral component of total urban India emissions (%)	Urban world baseline emissions (% of total)
1	Electricity	355.03	719.3	37.8	5.6	predominantly urban	719.3	100	59.26	8.6–13.0
2	Transport	80.28	142	7.5	4.5	87% is road based, 57.44% of total motor vehicles in the country owned by urban households	70.98	50.0	5.85	7.9–9.2
3	Residential[1]	78.89	137.8	7.2	4.4	urban households form 33.29% of total Indian households	45.88	33.3	3.78	4.7–5.5
4	Other energy[2]	78.93	100.9	5.3	1.9	petroleum refining is a peri-urban phenomenon	33.85	33.6	2.79	
5	Cement	60.87	129.9	6.8	6	predominantly urban	129.92	100.0	10.70	7.8–11.6

Table 13.1 (Continued)

S.No.	Sector	Million tons CO$_2$e 1994	Million tons CO$_2$e 2007	Sector contribution in total Indian GHG emissions (%)	CAGR (%)	Justifications to GHG contribution from the perspective of the location of activities that produced them	Urban India baseline emissions million tons CO$_2$e 2007	Sectoral emissions at national level (%)	Sectoral component of total urban India emissions (%)	Urban world baseline emissions (% of total)
6	Iron and steel	90.53	117.3	6.2	2	predominantly urban	117.32	100.0	9.66	
7	Other industry[3]	125.41	165.3	8.7	2.2	other metal and chemical is urban	38.92	23.5	3.21	
8	Agriculture	344.48	334.4	17.6	−0.2	non-urban	0	0.0	0.00	0
9	Waste	23.23	57.73	3.0	7.3	urban emissions considered in the assessment by INCAA	57.73	100.0	4.76	1.5
10	LULUCF	14.29	−177				0	0.0	0.00	
		1228.5	1728		2.9		1213.9	70.3	100.00	30.5–40.8

Source: The authors' estimates based on the data from 'India: Greenhouse Gas Emissions 2007', Indian Network for Climate Change Assessment, MoEF, India, 2010 and similar framework in Cities and Climate (UN Habitat, 2011) that is derived from Barker *et al.* (2007) and Satterthwaite (2008).

[1]Residential emissions also include the commercial sector

[2]Other energy includes petroleum refining and solid fuel manufacturing, agriculture and fisheries, fugitive emissions

[3]Other industries comprising pulp/paper, leather, textiles, food processing, mining and quarrying, and non-specific industries comprising rubber, plastic, watches, clocks, transport equipment, furniture, etc.

5.1   Electricity generation

Across the globe, fossil fuels are the major source of electricity. Globally, about two-thirds (67.1 per cent) of 20,181 TWh of electricity generated is sourced from thermal power plants (International Energy Agency, 2012), firing coal/peat, oil and petroleum gas as fuel and, thereby, creating 28,999.4 million tons of CO_2e emissions (International Energy Agency, 2011). In the case of India, there are over 100 thermal power plants collectively generating 138,806.18 MW of energy (Central Electricity Authority, 2012) and emitting 719.31 million tons of CO_2 (MoEF, 2010). The fuel mix is heavily dependent upon coal (90 per cent), followed by natural gas (8 per cent) and oil (2 per cent). Many thermal power plants are outside the municipal limits, but as per the modest definition of urban areas prevalent in India (discussed in Section 1), most of them qualify as being in towns, cities and urban agglomerations and almost the entire emission can be attributed to urban areas. This is drastically contrary to the world scenario where large fossil-fuel power stations are considered to be located outside the urban areas and contribute from 8.6 per cent to 13 per cent (Satterthwaite, 2008). The individual urban contribution in India is thus about 1.90 tons of CO_2e per capita annually.

5.2   Transport

As per the IPCC, globally, transportation contributes to about 23 per cent of the energy related to GHG emissions (IPCC, 2007). Cities thrive on mobility by means of air, sea, road and river-based modes. Historically, urban centres all over the world were located on natural gateways, trade highways and sea routes. By the middle of the 20th century, all of the 15 million-plus cities were seaports. In the last few decades, motorised transport has become the preferred choice and a significant polluter – about three-quarters of energy related GHG emissions came from road vehicles in 2004 (IPCC, 2007).

In India, transport is responsible for 7.5 per cent of the national emissions. This has grown at 4.5 per cent annually since a 1994 assessment to levels of 142.04 million tons of CO_2e in 2007 (see Table 13.1). Road transport accounts for the bulk of emissions (87 per cent). In the absence of any national accounting or survey on annual motor trips in cities, it is difficult to assess the urban contribution. The ownership of motor vehicles in urban households is thereby used as a substitute indicator to assess baseline emissions. As per the Census 2011 figures, 57.44 per cent or a little light motor vehicles) are owned by urban households. Hence urban areas contribute to about 71 million tons of CO_2e emissions. The average petroleum consumption is about 150 l/capita, as derived for 41 Indian cities (ICLEI, 2009). The average individual emissions are very low: 0.18 tons of CO_2e per capita against other global cities (e.g. London: 1.18; New York: 1.47; Los Angeles: 4.74; Kennedy *et al.*, 2009).

It is imperative to understand how transport-related emissions vary with changes to the built environment. An analysis of ICLEI data reveals that as cities expand from 40 to 400 sq. km, annual city fuel consumption increases threefold from 1,000,000 to 3,000,000 l, but the average fuel consumption diminishes from 300 to 125 l/capita, thereafter almost becoming constant. This suggests for an optimal size of modern cities planned for motor transport, from the point of view of low-carbon emissions.

Another reason for low per-capita emissions is the heavy use of non-motorised transport (NMT), such as bicycles, rickshaws and walking trips in many smaller cities. NMT trips range from 53 per cent to 36 per cent of total trips across cities with a population varying from 0.5 to 3.0 million, while the use of public transport ranges from 9 per cent to 22 per cent (MoUD, 2008). Hence it becomes significant to reduce emissions in this category of cities.

5.3 Residential and commercial

GHG emissions from residential and commercial buildings are caused by electricity use, lighting and illumination, space heating and cooling, direct emissions from cooking, embodied energy within building materials and so on. Globally, 10.6 billion tons of CO_2e or 8 per cent of GHG emissions are emitted in the sector (IPCC, 2007). In India, the sector contributes to 139.51 million tons of CO_2e or 7.5 per cent of national emissions. India's urban population resides in 110,139,853 households, which is about one-third (33.29 per cent) of total households (Census of India, 2011). As such, about 46.5 million tons of CO_2e is attributable to urban areas, with an average individual contribution of 0.12 tons of CO_2e per capita annually. It is significant to research how the emissions vary across different city types. A preliminary analysis of ICLEI (2009) data reveals that 37 out of 41 sampled Indian cities with electricity consumption of less than 2,000 million kW have total emissions of less than 5 million tons of CO_2e.

5.4 Industry

Direct GHG emissions from the industrial sector are about 7.2 gigatons of CO_2e, and total emissions, including indirect emissions, are about 12 gigatons of CO_2e in 2004, which constitutes 19 per cent of global GHG emissions (IPCC, 2007). These include the manufacture of iron and steel, metal processing, chemicals and fertilizers, cement production, pulp and paper, and are responsible for direct emissions. Depending upon the nature of activities, the contributions of industrial cities across the globe is substantially larger than for cities whose primary economic activities are non-industrial. It should be noted that it is the energy-intensive 'heavy' or 'manufacturing' industries that cause emissions, while many cities supporting an equally vibrant industrious workforce in cottage industries and finishing industries are less carbon emitting.

In India, cement, iron and steel, and other industries contributed to emissions of 129.92, 117.32 and 165.31 million tons of CO_2e in 2007, respectively (MoEF, 2010), collectively constituting 412.55 million tons of CO_2e or 22 per cent of national emissions. They have, respectively, grown at 6 per cent, 2 per cent and 2.2 per cent per annum since the last assessment in 1994. Most of the cement, and iron and steel industries in India are labour-intensive and, considering the modest definition of urban areas adopted in India, would be inappropriate to qualify as functioning as rural entities. As such, 247.24 million tons of CO_2e (i.e. $129.92 + 117.32$ million tons of CO_2e) can virtually be attributed to urban areas. On the same grounds, 38.92 million tons of CO_2e emitted from 'other metal and chemicals' in the 'other industries' category has urban associations. Hence a total of 286.16 million tons of CO_2e is produced in cities and forms a bulk of industrial emissions at 69.4 per cent and 16.6 per cent of the total nationwide emissions. This is comparable to the 7.9–9.2 per cent of GHGs allocated to cities (Satterthwaite, 2008). The per capita industrial energy consumption in urban area comes out to be 0.76 tons of CO_2e per annum.

5.5 Waste

Global GHG emissions from the waste sector are 1,300 million tons of CO_2e or 2.8 per cent of global GHG emissions (IPCC, 2007). Waste management in urban agglomerations becomes a critical issue of hygiene and sanitation. Indian emissions from the waste sector are of the order of 57.73 million tons of CO_2, forming 3 per cent of national emissions (MoEF, 2010). The methodology accounts only for urban-generated wastes, assuming that since waste in rural areas decomposes in disaggregated and aerobic conditions, it does not contribute to GHG emissions. Therefore the entire 57.73 million tons of CO_2 are urban in origin, with individual generation of 0.15 tons of CO_2 per capita.

India is still to achieve complete sanitation in urban areas as only 75–81 per cent of urban households have toilets while the average solid waste generation from 41 Indian cities is 0.61 kg/day (ICLEI, 2009). The annual generation and collection figures are likely to increase to 377 million tons and 295 million tons by 2030, thus widening the gap fourfold (Mckinsey, 2010). Reasons for low emissions from waste sector in India include limited methane gas recovery at landfill sites, incineration and waste-to-energy (WTE) projects installed (Central Pollution Control Board (CPCB), 2006).

5.6 Land use, land-use change and forestry

Globally, agriculture and forestry activities are allocated 13.5 per cent and 17.4 per cent of GHG emissions (IPCC, 2007). It is also recognised that agriculture and forestry activities also support the world's urban population regarding food grains, horticultural products, forest produce, non-timber forest produce and industrial raw materials that fuel the urban economy.

However, considering the location of activities from where the GHGs are produced, the LULUCF-associated emissions are essentially rural in origin. In India, this sector contributes to a net reduction of 177.03 million tons of CO_2 (MoEF, 2010). It notes that while the area of settlements/built-up lands has increased by 0.01 m ha in 2006–2007, The conversion in land use did not contribute to emission but a net removal of 0.04 million tons of CO_2. This requires appropriate scientific corroboration through recent satellite data, considering that land demand for various projects/schemes by the government itself during the 12th Five Year Plan (2007–2012) is 0.3 m ha (Sethi, 2011). According to certain estimates, the world's urban population is expected to double and the built-up area triple by 2030 (Angel *et al.*, 2005). This will pose enormous challenges to the mitigating capacities of agricultural lands and other global commons, such as forests, wetlands and lakes.

6. State of urban governance in India

Like many modern nation states, India is a democratic, sovereign republic with a federal governance structure bearing a centre-state system. The division of powers is defined in Article 245 of the Indian Constitution (available at http://india.gov.in/govt/constitutions_india.php, along with all of the amendments) under which 97 subjects of national importance – such as defence, foreign affairs, rail and highways. and interstate disputes – are under the Union List, 66 subjects are in the State List and 47 subjects are in the Concurrent List. It is worth noting that town planning, city planning, and civic or municipal management are not mentioned in any of these lists. Yet, the subjects relevant to cities have been interpreted from the state list (public health and sanitation, hospital and dispensaries, land, works, taxes on land and buildings) and from the concurrent list (economic and social planning) to further legislate in the field. While City Improvement Trusts (the first institutions established in modern India to manage civic functions) have existed in India since the 19th century, it was the model Town and Country Planning Law of 1960, prepared by the TCPO under the government of India, that led to various state acts, forming the basis to establish town and country planning offices in the states and districts across the country.

The enactment of the Delhi Development Act (1957) was followed by a spate in formation of development authorities for cities and special areas to further planned urban development. However, the roles and responsibility of development authorities in the execution of essential municipal functions was not clear. As such, municipal bodies were formed by states under respective state acts to perform basic civic functions. As per the most recent report of the Central Finance Commission (CFC), out of 7,935 census towns in India there are 3,842 ULBs in the form of municipal corporations, municipalities, municipal councils, municipal boards, *nagar panchayats*, town councils and notified area councils exercising authority, powers and functions delegated by the state (CFC, 2010).

In addition to the above, depending on certain specific needs of defence, industry, infrastructure, socioeconomics and shelter, centre or state governments constitute countless boards, authorities and trusts to plan or manage particular areas, such as cantonments, special economic zones, industrial estates, export regions or urban functions (e.g. city improvement, housing, shelter, heritage, power, water and sewerage). A sea change occurred in 1993 when, in order to decentralise power, the government conferred constitutional authority to the ULBs through the 74th Constitutional Amendment Act (CAA), making them pillars for Tier-III governance, below the tier of the centre and the states. This added Article 243 W (the 12th Schedule) defining 18 functions:

- urban planning including town planning;
- regulation of land use and construction of buildings;
- planning for economic and social development;
- roads and bridges;
- water supply for domestic, industrial and commercial purposes;
- public health, sanitation conservancy and solid-waste management;
- fire services;
- urban forestry, protection of the environment and promotion of ecological aspects;
- safeguarding the interests of weaker sections of society, including the mentally and physically disabled;
- slum improvement and upgrading;
- urban poverty alleviation;
- provision of urban amenities and facilities, such as parks, gardens and playgrounds;
- promotion of cultural, educational and aesthetic aspects;
- burials and burial grounds; cremations and cremation grounds; and electric crematoriums;
- cattle pounds; prevention of cruelty to animals;
- vital statistics, including registration of births and deaths
- public amenities, including streetlighting, parking lots, bus stops and public conveniences;
- regulation of slaughterhouses and tanneries.

As is evident, most of the above provisions have a considerable influence on the mitigating capacities of a city. But without actual functions, jurisdictions and so on devolved to them by line departments, parastatal authorities, boards or trusts already functioning in urban areas, the CAA remains a law in spirit only. Only ten states have transferred functions mandated under the 74th CAA to the ULBs (Jha and Vaidya, 2011). The act also provides for the constitution of the State Finance Commissions, every five years, to review financial administration by local governments and to suggest further transfers. However, as observed by the 13th CFC, there are long delays in the constitution and delivery of final reports of the respective state finance commissions.

Most of the urban bodies are weak and encounter immense operation and management pressure, staff shortages, a lack of sufficient capacities, equipment and technologies, and overlapping jurisdiction to deal with elementary urban issues (Jha and Vaidya, 2011; Sethi, 2011; Singh, 2011; Siwach, 2011). Expenditure on salaries and wages accounts for 54.2 per cent of the total municipal expenditure. In several states, however, it is as high as 80.4 per cent (India Infrastructure Report, 2006). In addition, there are external challenges such as the growing population, horizontal coordination, political allegiance and transparency, thus putting the mitigation of CC on the backburner.

CC finds a narrow reference in the national policy framework (Ministry of Environment and Forests, 2006). Most of the document clarifies India's position in the international CC debate rather than to offering a nationwide integrated approach on the subject. It upholds the principle of common but differentiated responsibilities and respective capabilities of different countries. Pertaining to mitigation, the policy emphasises multilateral approaches, rights to equal per-capita entitlements of global environmental resources, priority to the right to develop, and encouragement of Indian industry to participate in clean development mechanisms through capacity-building.

This is followed by the national-level plan adopted by central government, which identifies eight missions: National Solar Mission, National Mission for Enhanced Energy Efficiency, National Mission on Sustainable Habitat, National Water Mission, National Mission for Sustaining the Himalayan Ecosystem, National Mission for a Green India, National Mission for Sustainable Agriculture and National Mission on Strategic Knowledge for Climate Change (Ministry of Environment and Forests, 2008). Themes bearing strong potential to influence urban India are the national missions on sustainable habitat, energy efficiency, solar mission, green India and strategic knowledge for CC. There are two major concerns where the policy confines its outlook. First, it does not follow an integrated view but a squarely sectoral approach to containing GHG emissions. Second, it is limited to the identification of the institutional and procedural mechanisms that will enable the action plan to function, so as such it is virtually a vision paper.

Meanwhile, another major initiative from the urban perspective was introduced by national government in 2005 under the JNNURM to invest more than Rs1,000 billion in 65 cities for upgrading physical infrastructure, urban transport, housing for the urban poor and good governance. The funding was based on a strategic planning document called 'The City Development Plan'. It is striking that any aspect pertaining to CC mitigation or adaptation was not a prerequisite for this strategic vision. As in June 2012, 554 projects at a total cost of Rs622.53 billion have been sanctioned under the head of Urban Infrastructure and Governance (Press Information Bureau, 2012). It is gradually being understood that the mission has fallen short of building substantial technical and managerial capacities for ULBs, which is now being contemplated in the second phase of the JNNURM. Information technology including geographical information systems (GISs), municipal information systems (MISs) and e-governance will prove to be a cornerstone for the crumbling municipal bodies. As noted by the Second Administrative Reforms Commission (SARC), capacity-building measures should not be confined to the selected towns but should be available for all cities/towns (SARC, 2007). It also stresses that the entire practice of urban governance in India follows a top-to-bottom approach that CC mitigation mechanisms will need to acknowledge.

The fact of this case is that without any formidable power for the ULBs to collect tax and be financially sustainable, the current situation in India is bound to remain unchanged. The resource gap for JNNURM cities is estimated at Rs2,768.22 billion for 2005–2012 (13th CFC, 2010). The problem is acute in states where *octroi* has been abolished with no significant initiatives taken to expand the resource base (SARC, 2007; Thakur, 2011). As a study on property tax reforms in ten important metros of India reveals, four out of the ten sample cities determine tax on annual rateable value, with no market considerations, while only two cities revise the tax rate annually (Ghosh, 2010). This indicates limited leeway to push market-oriented mitigation mechanisms. Along similar lines to property tax reforms, the union government and the state finance commissions have examined the issue of user charges also and recommended that appropriate rates should be charged for the services provided by the municipal bodies. The union government has made it mandatory for ULBs in the JNNURM to levy reasonable user charges with the objective that the full cost of operation and maintenance is collected within the next five years (SARC, 2007). Thus any additional user charges on account of pushing GHG-mitigating technology and practices within the urban systems also need to be levied directly on users. This also offers a good opportunity to exercise the principle of 'polluter pays'. However, it should be highlighted that owing to the commitment to transparency, any additional carbon tax, cess or levy has to be deposited in a separate 'escrow' account or a 'fund' and be used to finance mitigation-related projects or capacity-building. There is also growing demand to use land banks, floor area ratio (FAR), transferable development rights (TDR) and accommodation reservation as resources to make ULBs independent (SARC, 2007; Chotani, 2011).

There are voices emanating internationally and from within the country that seek changes to the traditional role of ULBs – from provider to facilitator (SARC, 2007; Mckinsey, 2010; Jain, 2011; Siwach, 2011). It is gradually being realised that separation of the functions of policy, regulation and operations is necessary, and policy and regulatory functions can be with the government, and the delivery functions can be assigned to the private sector (SARC, 2007; Narendra, 2011). The role of citizens' participation could be enhanced once the powers are devolved to the proposed grassroots institutions, such as ward committees and area councils with well-defined functions (SARC, 2007). The Right to Information Act 2005 has tremendously empowered the common man to exercise informed choices. With concerted grooming, civil society holds immense potential in assimilating CC themes, such as awareness and education, resource and energy conservation, green buildings, waste reduction and greening cities.

7. Mitigation-inclusive urban governance framework

As Sheng (2010) abstractly explains, governance is 'the quality of the relationship between the government and the citizens' (p. 134). Operationally, it is defined as 'the quality of the process by which decisions are taken that affect public affairs, as well as the quality of the implementation and outcome of these decisions' (Sheng, 2010, p. 134). The mute question is how responsible urban governance can also lead to CC mitigation.

First, mitigation of CC in a city is relevant to many of activities and governance matters, such as public health, transport, land use, development controls, local taxes, waste management and public parks, which the local bodies are already committed to.

Second, there is growing recognition that investments in mitigation are particularly important in rapidly urbanising middle-income countries because long-lived capital stock, once established, can lock in emissions for long periods (potentially centuries) (World Bank, 2010).

Historically, cities have been places of social, economic and political innovation. Though in high-income nations many city officials have demonstrated a stronger commitment to GHG reduction than national politicians (Satterthwaite, 2011), the case for an urbanising India shows small and fragmented beginnings.

Yet, it has been observed that deficient intergovernmental relations, inadequate popular local representation processes, weak subnational institutions and poor financing mechanisms to support subnational government forms pose critical questions for various stakeholders (McCarney, 2012). If the challenges of CC and reduced GHG emissions are to be met, then it is essential that we understand how to reduce energy demands and their undesirable environmental impacts in cities (Lindfield, 2010). To reduce the energy demand in the operation of cities, it is essential that we understand the way in which urbanisation, city form, design, development density, logistics and urban-management systems can be made more efficient and effective.

In order to explore mitigation-inclusive urban governance, the most extensively mandated framework globally by UN Habitat and the OECD is followed. This reviews different modes of governance – that is, through self-governance, provision, regulation and enabling across various sectors, as discussed in detail below, and recommends suitable mitigation mechanisms therein (as summarised in Table 13.2).

7.1   Electricity generation

Electricity generation from fossil fuels is largely under the control of the National Thermal Power Corporation, and transmitted and regulated through the Power Grid Corporation and Central Electricity Agency, respectively, all being centralised public entities. The complete system is subject to centre-state negotiations with minimal representation of the ULBs. Other than limited mandate and finances on power, owing to direct carbon emissions and nuisance value, municipal governments find it difficult to even secure lands to install WTE plants within their jurisdiction, as seen in the case of Delhi Jindal Plant (Down to Earth, 2012). India has plans to enhance its thermal power-plant capacity significantly to produce 75,785 MW in the base-case scenario by 2017 (Government of India, 2012). Smart grids, power reforms to reduce generation, and transmission losses can prove to be good governance mechanisms. A reduction in fossil-fuel subsidies, taxes or carbon charges on fossil fuels, renewable energy obligations and loans on renewable energy production are some international best practices (World Bank, 2009).

An additional consideration of 18,500 MW kept in the base-case scenario for 2017 through renewable sources will be a major thrust to reduce emissions. The devolution of distribution functions to ULBs or regulatory powers can strengthen their position. The cap and trade mechanism between states where ULBs also have a stake will trigger low-carbon technologies and practices, and complete the demand and supply side of the sector.

7.2   Transport

Urban local bodies play a significant role in the mitigation of carbon emissions arising from motor transport. In many Indian cities there are sector-specific parastatal agencies such as the transport corporation/undertaking to manage city traffic and bus services. The case of the capital city has seen a reduction in direct CO_2 emissions from public transport after the fuel switch to natural gas, and as the Central Road Research Institute study showed, Delhi Metro saved 16,60,000 vehicle km and 2,275 tons of GHG emissions by 2007 (SARC, 2007). Many metropolitans, such as Ahmedabad, Bangalore, Kolkata, Mumbai, Kochi, Jaipur and Pimpri Chinchwad, are

Table 13.2 Mitigation-inclusive urban governance framework in India – summary table

	Self-governing	Regulation	Provision	Enabling
Energy generation	energy audits and planning for all municipal services; energy saved is energy produced installation of model WTE projects by ULBs enhance purchased renewable energy purchase of energy-efficient appliances and equipment in public infrastructure	setting up of cap and trade mechanism between states, where cities can also trade	transfer of electricity functions to ULBs or make distribution companies accountable provision of smart grid service for consumers to link them to active demand–supply situation	incentives for projects to be installed for a ULB in city limits or near to captive plant on renewable energy, such as solar, wind, WTE campaigns for energy conservation
Transport	GIS and GPS system to monitor all municipal vehicles reduction of emissions from employees, through mobility plan, car pools, etc. procurement of low-carbon-emitting vehicles – BS IV compliant for city-fleet and municipal vehicles demonstration projects to economically produce biofuel use of low-carbon technology in road construction and maintenance overhaul	setting up of urban metropolitan authorities planning and regulations for transit-oriented development congestion charges and high parking prices in congested areas removal of tax and subsidies on fuels oil companies to supply biofuels in municipal areas high taxation on personal vehicle registration, especially on inefficient or luxury models	greater percolation of mass rapid transport, especially electricity based provision of infrastructure for alternative or non-motorised modes, such as cycle, walk, rickshaw	incentives, tax relaxations on use of hybrid and electric vehicles awareness and education campaigns to use low-carbon or zero-emission transport modes incentives on FAR for transit-oriented development projects

Table 13.2 (Continued)

	Self-governing	Regulation	Provision	Enabling
Residential and commercial	switch to green buildings and energy-efficient technologies in all new, renovation of municipal buildings demonstration projects on house, neighbourhood, community scales for replication	revision of norms, standards for higher FAR, TDR in municipal core to encourage compact habitations and mixed land use penalties or high property tax on vacant houses to curb speculative demand mandatory application of ECBC throughout India all-new residential and commercial developments to conform to green building guidelines	transfer of planning powers and provisional functions to ULBs heavy investments in rental housing, labour housing, studio apartments, industrial housing, which is low cost and low carbon footprint provision of low-carbon municipal infrastructure	rebates on property taxes for green buildings, mixed-use development guidance to architects, engineers, developers on energy-efficient buildings campaigns for green buildings ease of grants/loans for low-carbon technologies in households and business
Industry	low-carbon considerations in management of ULB's estates, procurements, etc.	transfer of regulatory functions of SPCB to ULBs or industrial estates in urban limits cap and trade market with reduction targets for GHG emissions polluter pays penalties on high emissions – chimney and captive plants/diesel generator sets	phased overhaul of municipal infrastructure in industrial estates provision of designated power supply to reduce load on captive (diesel generator) generation	ease in grants/loans to industries for technology overhaul enable list of energy auditors and firms incentives of higher FAR, TDR for low emissions rebates on property tax for consistent compliance

Table 13.2 (Continued)

	Self-governing	Regulation	Provision	Enabling
Waste	waste prevention, recycle, reuse within the ULB procurement of low-carbon waste technologies and management practices local solutions for waste segregation, composting and incineration model projects for scientific landfill, methane recovery and WTE use of GPS, GIS and MIS to measure and monitor waste generation, collection, transport, composting, dumping and incineration activities	ban on excessive packaging and plastic products penalties on non-segregated waste at household or community level regulation of municipal and e-waste through welfare associations and accredited junk dealers	recycling, recomposting, reuse schemes by ULBs overhaul of waste water infrastructure and full coverage in municipal areas	campaigns for reusing, recycling, reducing waste promote use of recycled products enable greater role of private/PPP, NGOs and citizen groups in solid-waste management awareness campaigns against direct combustion of compostable waste
LULUCF	plantation in city parks, gardens, public places, city forests with self-efforts or joint teams with forest department, NGOs, welfare associations, schoolchildren, etc.	planning and regulatory functions to revise masterplan against horizontal sprawl and suburbanisation transfer of regulatory functions related to planning, change of land-use, development controls to ULBs in municipal areas and peripheral areas higher FAR to urban cores ensuring mixed land use linking plantations with cap and trade market of industrial and power sector setting up of minimum standards for keeping public parks and large private properties under green cover	devolution of management rights of parks, playgrounds and gardens towards committees and welfare associations	incentives for peripheral properties to retain agricultural land use awareness campaigns for enhancing tree cover

Source: Author's recommendations, adapted from internationally followed frameworks by UN Habitat (2011) and OECD (2009), derived from Bulkeley and Kern (2006), Bulkeley *et al.* (2009), Martinot *et al.* (2009) and ICLEI (2010).

gradually shifting to mass-transit metros, bus-rapid transit, light rail systems and so on.

Further emphasis is needed on mandatory fuel economy, biofuel blending and CO_2 standards for road transport, taxes on vehicle purchase, registration, motor fuels or roads, parking pricing, and investment in public and land-use integrated transport (World Bank, 2009; Ahmed and Choi, 2010). Integrated land-use planning with transit corridors by assigning development hubs and allocating incentive zoning as executed in Sau Paulo, Hong Kong and Shanghai can result in a positive shift towards public transport and, thereby, reducing carbon emissions, as against indiscriminately permitting mixed land-use along the entire corridor as proposed in the Master Plan of Delhi. The national guidelines on this issue recommend setting up Urban Transport Authorities/Unified Metropolitan Transport Authorities in million-plus cities in India (MoUD, 2006). At present, such authorities have been constituted in a couple of metropolitans only, and their effectiveness has to be critically reviewed at the cost of not jeopardising the authority of the ULBs.

7.3　Residential and commercial

Since emissions in the residential and commercial sector are associated with broad energy needs for sustenance and socioeconomic well-being, these are representative of the built habitat an urban society lives in, and also how cities chose to plan and govern themselves. In most of cities worldwide, residential and commercial land use constitutes more than half of the settlement. Emissions in this sector in India are on the rise and show large urban–rural divide (Gupta and Chandiwala, 2009; Parikh and Parikh, 2011). In terms of lifestyle differences across household expenditure classes, the urban top 10 per cent accounts for emissions of 4,099 kg per capita per year, while the rural bottom 10 per cent accounts for only 150 kg per capita per year (Parikh *et al.*, 2009). In the context of cities, it becomes imperative to assess these emissions with respect to factors that shape the built environment – namely, city size (in terms of both population and area), city form (population density) and geographical location.

Indian cities in the last decade have grown by 44 per cent from natural population growth, and 56 per cent from rural–urban migration and changes in boundary definitions (TCPO, 2012). In the absence of any policy or regulatory mechanism to permit or limit the movement of people across state or local jurisdictions, the municipal bodies in India have no power to control migration. Analysis of selected Indian cities for which data are available suggests that as the municipal area expands from 40 sq. km to 550 sq. km, the overall city emissions rise from 1 million tons of CO_2 to over 7 million tons of CO_2, thereby, showing a strong correlation between city size and fuel emissions (ICLEI, 2009). Most Indian cities with a population above 0.5 million have development authorities (parastatals) established under the act of legislature and conferred powers to delineate planning area, formulate development plans, propose land-use plans and develop controls. Municipal authorities are conferred civic management, sanitation, provision of basic amenities within its jurisdiction area. Often this includes sanctions of building permissions, but otherwise ULBs have not been devolved powers to determine the city area or the built form through scientific land-use planning and development controls.

As the density increases there is a concentration of services, and people use more public and non-motorised forms of transport, reducing the transportation energy use per capita. Good land-use policies can encourage this trend. European cities tend to be more compact with greater reliance on public transport (Dodman, 2009). The adjacency of residential and commercial areas and mixed land-use development are central to this theme. The overall availability of land per person in India has declined steadily, from 1.28 hectares per person in 1901 to 0.32 hectares in 2001 (Ribeiro, n.d.). Analysis of ICLEI data for 41 Indian cities shows that as city density increases from 2,500 to 25,000 per sq. km, carbon emissions rise from 1 to 7.5 million tons of CO_2 while per capita carbon emissions decline from 1 to 0.75 tons of CO_2, thus showing benefits on account of agglomeration and the sharing of infrastructure. Paradoxically, density is hardly exercised as a planning tool by authorities in India to shape cohesive and compact habitations. It is surprising to see that even JNNURM, essentially an urban renewal programme of national significance, has hardly contributed to brownfield development.

Allocating higher FAR and tradable development rights, moderating property tax in inner areas can result in counter sprawl. Additionally, residential and commercial structures need to be made smarter through energy-efficient devices, green building practice and waste-management systems to reduce community and municipal emissions. The Green Rating for Integrated Habitat Assessment (GRIHA), developed jointly by the Energy and Resources Institute and the Ministry of New and Renewable Energy of the government of India, is a green building 'design evaluation system' and is suitable for all kinds of building in different climatic zones of the country. According to GRIHA it is estimated that the average green building reduce energy use by more than 30 per cent, consumes less water and limits the waste sent to landfill sites (GRIHA, 2012).

Speculative investments made by non-genuine buyers are evident that lock in infrastructure and embodied energies. As reported recently, 10.73 per cent of houses in India are either vacant or locked, and most of this is under the fresh housing stock created in India during the last decade (Ministry of Housing & Urban Poverty Alleviation, 2012). The local authority needs to reconsider its age-old view on housing and should invigorate schemes for rental housing, industrial housing, studio apartments and transient housing across the socioeconomic spectrum. The urgent need to regulate the real-estate sector has been recommended time and again (SARC, 2007; Revi, 2008) and will go a long way in achieving low-carbon habitations.

International discourse mandates appliance standards and labelling, building codes and certification, demand-side management programmes, public-sector leadership programmes including e-procurement, and incentives for energy service companies (World Bank, 2009). The Indian government recognises that, for good governance, municipal bodies should be encouraged to take responsibility for urban power distribution in their areas. This, however, should be done after adequate capacity-building in these organisations (SARC, 2007). Further, municipal building bylaws should incorporate power-conservation measures. In this regard, mandatory application of the Energy Conservation Building Code (ECBC) all over India is still pending.

7.4 Industry

With the emergence of the tertiary (services) sector in the last few decades, a reverse trend of disagglomeration of industries to the periphery or outside the city limits has started. In the large metropolitan cities, local government is under constant pressure of judiciary, citizen groups, non-governmental organisations (NGOs) and so on to provide a clean and green environment for the growing middle-income majority, which is educated, environment conscious and earns from the service industries. As such the ground reality of the sector from a governance and policy perspective is quite complex and confusing. Many cities housing 'heavy industry', such as iron and steel plants, locomotives and cement, such as Bhilai, Tatanagar and Rourkela, are governed by the industrial authorities themselves. Lately, Special Economic Zones have also been empowered for planning, provisioning infrastructure and maintenance, but not with monitoring ambient air quality or carbon emissions, because environmental functions are still with the respective state pollution-control boards. This reflects a limited and fragmented hold of the government, and calls for complete devolution of managerial powers upon ULBs.

Global best practices recommend provision of benchmark information, performance standards, subsidies, tax credits, tradable permits and voluntary agreements (World Bank, 2009). The first ever Emission Trading System was initiated in Tokyo in 2010 (PADECO report cited in UN Habitat, 2011) with a target to reduce CO_2 emissions by 25 per cent below 2000 levels by 2020. Like power generation, it is high time that industries in urban areas are regulated by a cap and trade market mechanism with the government and citizen groups collectively acting as watchdogs. Measurement and monitoring by the ULBs will be a fundamental step in this direction.

7.5 Waste

Urban local bodies worldwide and in India have been constituted to provide basic services of sanitation. The very municipal act bestows the onus of managing city sewers, waterlogging and garbage on these authorities. In India, this has been strengthened by the Municipal Solid Waste Rules 2000 under the Environmental Protection Act 1986, by which the responsibility to manage solid waste by scientific planning and management, and the setting up of landfill sites and treatment plants, is delegated to the municipal bodies. However, as far as capacities are concerned, the municipal authorities have limited human, material and financial resources to scientifically manage waste. It is surprising to note that many local bodies lack the basic knowledge and equipment to measure and handle waste. Though the generation of waste is increasing, it is difficult to agree that mitigation capacities are developing at the same pace.

Global think tanks suggest financial incentives for improved waste and waste-water management, renewable energy incentives or obligations, and waste-management regulations (World Bank, 2009). Effective waste management is dependent on how well cities garner alliances between diverse stakeholders who are active in the field, such as private entrepreneurs, NGOs, citizen groups and the municipal government itself. The carbon reclaimed in the WTE practices, though small, will give fourfold benefits by halving the GHG emissions from waste while doubling energy production from the present levels, so this promises significant opportunities in the future.

The SARC recommendations for solid waste management issues in local governance mandates that (i) in all towns and cities with a population above 0.1 million the possibility of taking up public–private partnership (PPP) projects for the collection and disposal of garbage may be explored; after (ii) creating suitable capacities to manage contracts; (iii) municipal bylaws/rules should provide for the segregation of waste into definite categories based on its manner of final disposal; and (iv) special solid-waste management charges should be levied on units generating large amounts of solid waste. Thus a balanced, enabling and regulatory approach seems most appropriate for the Indian case.

7.6 Land use, land-use change and forestry

The sector has been undermined for urban forestry and agriculture, though many cities have intermittent yet significant open lands available. The governance is generally not supportive to actually enhance tree cover or promote rooftop agriculture/gardening as seen in many European settlements. Yet vast potential exists: as per the recent figures, green cover in Delhi has increased from 26 sq. km in 1997 to about 296.20 sq. km in 2009 (Forest Survey of India, 2011) as a result of an eco joint force being set up. This again offers significant potential as it enhances the mitigation capacities of the urban system, reduces the heat-island effect and thereby limits the use of air-conditioning in buildings and vehicles, and also diminishes transportation emissions resulting from the haulage of food into cities from villages. The augmentation of the assimilative capacities to local carbon emissions creates additional benefits. The small but incremental land-use changes for green areas within a city and their impact on enhancing mitigation and assimilation capacities as a whole need further empirical research.

Statutory bodies involved in plan preparation have been indulgent in acquiring cheaply available lands for the long-term planning process and have thereby driven suburbanisation, horizontal sprawls and change in land use and land cover. There is no incentive to citizens to save their land for environmental provisions and to go vertical, to reduce unnecessary travel time and to enjoy a better lifestyle. As Indian cities grow from 100, 500, 1,000, 2,000 to 3,000 sq. km, the average trip length increases from 3, 6.5, 7, 10 to 12 km, respectively (MoUD, 2008).

Depending upon city size, the policy for any new planned development seeks 12–25 per cent under recreation (Urban Development Plan and Formulation Guidelines India, 1996), and many planned cities do have sufficient recreation areas with a multitier hierarchy of open spaces. As such, there is much potential through green management to mitigate carbon emissions if tackled earnestly. Lower opportunities prevail in cities where the planning authorities were unable to meet the development pace, and have let squatter settlements, slums and unplanned colonies to come up, with no local parks and gardens.

Meanwhile, ULBs that manage the city core are already under acute pressure to save precious vacant plots for municipal and sociocultural functions and find urban greenery as a financial liability rather than an opportunity. Recently, metropolitan cities like Delhi have shown the way and devolved municipal functions to maintain city parks to local wards and welfare associations. However, this is more to reduce the responsibility of the government than to empower grassroots institutions. There is much deliberation about actually making urban green management a conscious city-wide effort, participatory and outcome-oriented. Additionally, all such functions need to be made financially sustainable by integrating them with taxes or market mechanisms that are needed in the power and industry sectors. Thus there is a far greater role to be played by ULBs in this sector through regulation, specifically in active land-use planning.

8. Limitations

The research deals with CC mitigation in urban areas by focusing upon contributions and drivers of GHG emissions. The most recently recorded GHG data (2007) published by the Indian government have been utilised to explore a baseline urban scenario. A revised release from the authorities will further assist in appropriating the results. The method of downscaling national GHG emissions is a componentwise derivative approach – essentially top-down analysis. It is open to validation by a bottom-up approach involving additive methods to assimilate GHG emissions for different hierarchies of urban areas in the country.

9. Conclusion

The research reveals that in the light of economic development, growing urbanisation and GHG emissions, the relevance of cities and their governance mechanisms cannot be undermined. There are definite drawbacks in prevailing governance mechanisms and ill-preparedness to mitigate CC. While urban areas contribute to almost two-thirds of national GDP and 70.3 per cent of national GHG emissions, they have limited resources – merely 1.59 per cent of GDP as total expenditure on urban infrastructure.

Further, the capacities of ULBs are deeply imperilled on account of the devolution of powers. While electricity generation from thermal power plants is the chief contributor of emissions (about 60 per cent), followed by the cement, and iron and steel industries (about 10 per cent each), the ULBs have no or restricted mandate over these subjects. In spite of the clear directions of the 74th CAA and recommendations made by various commissions, the states have done little to provide local autonomy to the ULBs to deal with sources or activities that contribute to GHG emissions and hence influence CC. The government's policy on the subject lacks sectoral integration, financing and market-based instruments to control GHG emissions. Mitigation of CC is also absent in the biggest government investments and reforms scheme in the urban sector – the JNNURM. Nonetheless, mitigation of CC provides a good opportunity for cities to revisit their energy needs, regulations, provisions and enabling mechanisms, and also to review waste-management and land-use/land-cover management practices. It is also imperative to further research on urban-associated GHG emissions from the perspective of the consumption pattern.

Application of an international mitigation framework to urban governance in India indicates that the steps to be taken vary drastically across sectors. They need to be market oriented with trading systems, where the government acts as the regulator only, as in the case of industry, power and LULUCF to a mix of authoritative and provision based measures (along with the private sector and civil society organisations), as in case of the transport and waste sector, or a blend of restricting and enabling mechanisms that are required to control indiscriminate urban sprawl. Yet, one aspect is common: that all sectors require equally strong communication, participation and incentives to stakeholders to move forward on low-carbon pathways.

Before outreach and collaborations, there are numerous initiatives to be taken by the government to set its house in order. The future of urban GHG emissions in India depends hugely on how well various levels of the federal system – centre, states and local bodies – respond to the call on CC, and exercise powers and share collective responsibilities among themselves. It is an appropriate time for the ULBs to stand up for innovative regulatory and market-based governance, while the states devolve larger city-planning and management functions to these constitutional bodies, and the national-level government responds to a sharpening of policy, mustering international support and financial commitments for local-level action.

The research, which is the first of its kind, presents the baseline situation of CC mitigation in an increasingly urbanising India, putting a strong emphasis on estimating and managing the GHG burden for appropriate urban governance. It offers an essential base to further investigate the results in light of any new GHG accounting released by the government, preferably at the sub-national or local levels. The results obtained from the downscaling method can be validated by appropriate incremental methods upscaling 'per city' or 'per capita' emissions, especially from ICLEI-like city inventories. It would also be interesting to study how city GHG emissions vary with a variation in physical or spatial parameters, such as city scale, form, geographical location, agglomeration, transport network, predominant land use and waste generation. The urban governance framework can be further expanded to include requirements for CC adaptation.

Abbreviations

BS IV	Bharat Standards – IV
CAA	Constitutional Amendment Act
CAGR	Cumulative Aggregated Growth Rate
CC	climate change
CFC	Central Finance Commission
CO_2	carbon dioxide
CO_2e	carbon dioxide equivalent
CPCB	Central Pollution Control Board
ECBC	Energy Conservation Building Code
FAR	floor area ratio
GDP	gross domestic product
GHG	greenhouse gas
GIS	Geographical Information System
GPS	global positioning system
GRIHA	Green Rating for Integrated Habitat Assessment
HPEC	High Powered Expert Committee on Urban Infrastructure
ICLEI	International Council for Local Environmental Initiatives
IIED	Institute of Environment & Development
INCCA	Indian Network for Climate Change Assessment
IPCC	Intergovernmental Panel on Climate Change
JNNURM	Jawaharlal Nehru Urban Renewal Mission
LULUCF	land use, land-use change and forestry
MIS	municipal information system
MoEF	Ministry of Environment & Forests

MoUD	Ministry of Urban Development
NGO	non-governmental organisation
NMT	non-motorised transport
OECD	Organisation for Economic Co-operation and Development
PPP	public–private partnership
SARC	Second Administrative Reforms Commission
SPCB	State Pollution Control Board
TCPO	Town and Country Planning Organisation
TDR	transferable development rights
ULB	Urban Local Bodies
UN	United Nations
UN DESA	United Nations Department of Economic and Social Affairs
UNFCCC	United Nations Framework Convention on Climate Change
UNFPA	UN Population Fund
WTE	waste to energy

References

S. Ahmed and M.J. Choi (2010) 'Urban India and Climate Change Mitigation Strategies towards Inclusive Growth', *Theoretical and Empirical Research in Urban management*, 6, 15, 60–73.

S. Angel, S.C. Sheppard and D.L. Civco (2005) *The Dynamics of Global Urban Expansion* (Washington DC: World Bank), p ii.

Barker T., I. Bashmakov, L. Bernstein, J. E. Bogner, P. R. Bosch, R. Dave, O. R. Davidson, B. S. Fisher, S. Gupta, K. Halsnæs, G. J. Heij, S. Kahn Ribeiro, S. Kobayashi, M. D. Levine, D. L. Martino, O. Masera, B. Metz, L. A. Meyer, G.-J. Nabuurs, A. Najam, N. Nakicenovic, H.-H. Rogner, J. Roy, J. Sathaye, R. Schock, P. Shukla, R. E. H. Sims, P. Smith, D. A. Tirpak, D. Urge-Vorsatz and D. Zhou (2007) 'Technical summary', in B. Metz, O. R. Davidson, P. R. Bosch, R. Dave and L. A. Meyer (eds) Climate Change 2007: Mitigation, Contribution of Working Group III to the Fourth Assessment Report of the Intergovernmental Panel on Climate Change,(Cambridge and New York: Cambridge University Press) 25–93, http://www.ipcc.ch/pdf/assessment-report/ar4/wg3/ar4_wg3_full_report.pdf, last accessed 7 October 2012.

H. Bulkeley and K. Kern (2006) 'Local Government and the Governing of Climate Change in Germany and the UK', *Urban Studies*, 43, 12, 2237–2259.

H. Bulkeley, H. Schroeder, K. Janda, J. Zhao, A. Armstrong, S.Y. Chu and S. Ghosh (2009) 'Cities and Climate Change: The Role of Institutions, Governance and Urban Planning', Paper prepared for the Fifth Urban Research Symposium, 'Cities and Climate Change: Responding to an Urgent Agenda', June 28–30, Marseille, France.

Census of India (2011) 'Provisional Population Totals -2011', Paper–II, Vol– II (New Delhi: Census of India), p. 1.

Central Electricity Authority (2012) Renovation, Modernisation and Life Extension of Thermal Power Stations, Quarterly Review Report (July–September 2012), http://www.cea.nic.in/, date accessed 15 September 2012.

CPCB (2006) *Assessment of Status of Municipal Solid Waste Management in Metro Cities and State Capitals* (New Delhi: Central Pollution Control Board), pp. 16–17.

CFC (2010) *'Local Governance' in Report of Thirteenth Central Finance Commission* (New Delhi: Government of India), pp. 149–185.

M.L. Chotani (2011) 'Managing Urban Land: Legislative and Planning Measures' in Listed papers of 59th National Town and Country Planning Congress, Panchkula, Haryana on Land as Resource for Urban Development, pp. 294–302.

D. Dodman (2009) 'Urban Form, Greenhouse Gas Emissions and Climate Vulnerability', in J.M. Guzman, G. Martine, G. McGranahan, D. Schensul and C. Tacoli (eds.) *Population Dynamics and Climate Change* (New York: UNFPA; London: IIED), pp. 64–79.

Down to Earth (2012) *Homepage* (New Delhi: Centre for Science and Environment).

Forest Survey of India (2011) *India Report* (New Delhi: Ministry of Environment and Forests), pp. 118–121.

D. Ghosh (2010) 'Property Tax Reforms in India', *Urban India*, 30, 2, 66–86.

Government of India (2012) *Report of The Working Group on Power for Twelfth Plan 2012–17* (New Delhi: Ministry of Power), p. 14.

GRIHA (2012) Green Rating for Integrated Habitat Assessment, http://www.grihaindia .org/, date accessed on 18 September 2012.

R. Gupta and S. Chandiwala (2009) 'A Critical and Comparative Evaluation of Approaches and Policies to Measure, Benchmark, Reduce and Manage CO2 Emissions from Energy Use in the Existing Building Stock of Developed and Rapidly-Developing Countries – Case Studies of UK, USA, and India' Paper presented at the Fifth Urban Research Symposium 2009, 'Cities and Climate Change: Responding to an Urgent Agenda', Marseille, June 28–30, p. 22.

HPEC (2011) *Report on Indian Urban Infrastructure and Services* (New Delhi: The High Powered Expert Committee on Urban Infrastructure, Government of India), p. xxvi–3.

ICLEI (2009) *Energy and Carbon Emissions Profiles of 54 South Asian Cities* (New Delhi: International Council for Local Environmental Initiatives).

ICLEI (2010) Cities in a Post-2012 Climate Policy Framework: Climate Financing for City Development? Views from Local Governments, Experts and Businesses, ICLEI, Bonn, www.iclei.org/fileadmin/user_upload/documents/Global/Services/Cities _in_a_Post-2012_ Policy_Framework-Climate_Financing_for_City_Development _ ICLEI_2010.pdf, last accessed 14 October 2010.

India Infrastructure Report (2006) *'Urban Infrastructure' in India Infrastructure Report* (New Delhi: Indian Development & Finance Corporation).

International Energy Agency (2011) *CO2 Emissions from Fuel Combustion, Ed 2011* (Paris: IEA/OECD), p. 46.

International Energy Agency (2012) Key World Energy Statistics-2011, www.iea.org accessed on 14 August 2012, 6–57.

IPCC (2007) *Climate Change 2007: Mitigation. Contribution of Working Group III to the Fourth Assessment Report of the Intergovernmental Panel on Climate Change* (Cambridge and New York: Cambridge University Press), p. 29.

A.K. Jain (2011) 'Urban Challenges for India and Revisiting Urban Planning', *Nagarlok*, XLIII, 3, 39–50.

G. Jha and C. Vaidya (2011) 'Role of Urban Local Bodies in Service Provision and Urban Economic Development: Issues in Strengthening Their Institutional Capabilities', *Nagarlok*, XLIII, 3, 1–38.

C.A. Kennedy, A. Ramaswami, S. Carney and S Dhakal (2009) 'Greenhouse Gas Emission Baselines for Global Cities and Metropolitan Regions', Paper presented at the Fifth Urban Research Symposium 2009, 'Cities and Climate Change: Responding to an Urgent Agenda', Marseille, p. 28.

M. Lindfield (2010) 'Cities: A Global Threat and a Missed Opportunity for Climate Change', *Environment & Urbanisation Asia*, 1, 2, 105–129.

E. Martinot, M. Zimmerman, M.V. Staden and N. Yamashita (2009) 'Global Status Report on Local Renewable Energy Policies', June 12 working draft, Collaborative report by REN21 Renewable Energy Policy Network, Institute for Sustainable Energy Policies (ISEP) and ICLEI Local Governments for Sustainability, http://www.ren21 .net/pdf/ REN21_LRE2009_Jun12.pdf, last accessed 15 October 2010.

P.L. McCarney (2012) 'City Indicators on Climate Change: Implications for Governance', *Environment & Urbanisation Asia*, 3, 1, 1–39.

Mckinsey (2010) *India's Urban Awakening: Building Inclusive Cities Sustaining Economic Growth* (New Delhi: Mckinsey Global Institute). pp. 13–35.

Ministry of Housing and Urban Poverty Alleviation (2012) *Report of the Technical Group on Urban Housing Shortage (TG-12), 2012–17* (New Delhi: Government of India), p. 14.

MoEF (2006) *National Environment Policy* (New Delhi: Ministry of Environment and Forests, Government of India), pp. 41–43.

MoEF (2008) *National Action Plan for Climate Change* (New Delhi: Ministry of Environment and Forests, Government of India), pp. 4–7.

MoEF (2009) *India's GHG Emissions Profile* (New Delhi: Climate Modelling Forum), p. 9.

MoEF (2010) *India: Greenhouse Gas Emissions 2007* (New Delhi: Indian Network for Climate Change Assessment, Ministry of Environment & Forests), pp. 1–54.

MoUD (2006) *National Urban Transportation Policy* (New Delhi: Ministry of Urban Development).

MoUD (2008) *Study on Traffic and Transportation Policies and Strategies in Urban Areas in India* (New Delhi: Ministry of Urban Development), p. 67.

MoUD (2011) *India's Urban Demographic Transition: The 2011 Census Results-Provisional* (New Delhi: JNNURM Directorate and National Institute of Urban Affairs), pp. 2–4.

A. Narendra (2011) 'Regulatory Framework for Urban Services in India', *Nagarlok*, XLIII, 3, 51–62.

OECD (2009) *Cities, Climate Change and Multilevel Governance*, OECD Environmental Working Papers No 14 (Paris: OECD Publishing), pp. 30–44.

J. Parikh, M. Panda, A. Ganesh-Kumar and V. Singh (2009) 'CO$_2$ Emissions Structure of Indian Economy', *Energy*, http://www.irade.org, date accessed 18 June 2012.

J. Parikh and K. Parikh (2011) 'India's Energy Needs and Low Carbon Options', *Energy*, http://www.irade.org, date accessed 18 June 2012.

Press Information Bureau (2012) 'JNNURM Triggers Investments in Urban Infrastructure Sector', http://pib.nic.in released on 11 October 2012, date accessed on 15 October 2012.

A.Revi (2008) 'Climate Change Risk: An Adaptation and Mitigation Agenda for Indian Cities', *Environment and Urbanisation*, 20, 1, 207–229.

Ribeiro (n.d.) 'Urban Local Bodies and Decentralised Planning' in Second Administrative Reforms Commission Report, p. 297.

SARC (2007) *6th Report: Local Governance* (New Delhi: Government of India), pp. 198–309.

D. Satterthwaite (2008) 'Cities' Contribution to Global Warming: Notes on the Allocation of Greenhouse Gas Emissions', *Environment & Urbanisation*, 20, 2, 539–549.

D. Satterthwaite (2009) 'The Implications of Population Growth and Urbanisation for Climate Change', in J.M. Guzman, G. Martine, G. McGranahan, D. Schensul and C. Tacoli (eds.) *Population Dynamics and Climate Change* (New York: UNFPA; London: IIED), pp. 45–63.

D. Satterthwaite (2011) 'How Urban Societies Can Adapt to Resource Shortage and Climate Change', *The Royal Society A*, 369, 1762–1783.

M. Sethi (2011) 'Alternative Perspective for Land Planning in Today's Context-Minimising Compensation Issues', *Indian Valuer*, XLIII, 4, 443–448.

Y.P. Sheng (2010) 'Good Urban Governance in Southeast Asia', *Environment & Urbanisation Asia*, 1, 2, 131–147.

S. Singh (2011) 'Urban Governance and the Role of Citizen Participation – A Case of Small and Medium Towns of India', *Urban India*, 31, 1, 138–153.

R. Siwach (2011) 'Capacity Building of Urban Local Bodies for Urban Development: Some Emerging Areas for NGO Intervention', *Urban India*, 31, 1, 15–31.

TCPO (2012) *Data Highlights (Urban) Based on Census of India* (New Delhi: Town & Country Planning Organisation), pp. 1–19.

S. Thakur (2011) 'Evaluating the Financial Performance of Indian Cities: A Diagnostic Report', *Urban India*, 31, 2, 90–110.

UN DESA (2011) *Urban Population Development and the Environment Wallchart* (New York: UN).

UN Habitat (2011) *Cities and Climate Change* (1st ed.) (London and Washington DC: Earthscan), pp. 10–33.

UNFCCC (1992) United Nations Framework Convention on Climate Change – Article 4, http://unfccc.int/essential_background/convention/background/items/1362.php accessed on 5 October 2012.

Urban Development Plan and Formulation Guidelines India (1996) (New Delhi: Ministry of Urban Development & Poverty Alleviation), p. 147.

World Bank (2009) *Climate Resilient Cities – A Primer on Reducing Vulnerabilities to Disasters* (Washington D.C.: World Bank), p. 18.

World Bank (2010) *Cities and Climate Change: An Urgent Agenda* (Washington D.C.: World Bank), pp. 14–32.

14
Unripe Fruits or Non-Raining Clouds? Climate Change Governance and the Funding Dilemma in Nepal

Bimal Raj Regmi and Dinanath Bhandari

1. Introduction

Climate change (CC) is one of the major environmental problems challenging society and the ecological system (Huq and Reid, 2004). The negative consequences of industrialisation and environmental degradation have increased the amount of greenhouse gases contributing to global warming and other associated problems. Global warming has led to the unanticipated variability and changes in the weather and climatic cycles, leading to accelerated changes and impacts. These climatic variations are threatening food security, ecosystem integrity and social harmony (Ayers, Alam and Huq, 2010), and the impact is hitting hardest the populations in low-income and developing countries.

Nepal is considered to be one of the most vulnerable countries in the world due to its fragile climate-sensitive ecosystem and socioeconomic circumstances (Regmi and Adhikari, 2007). According to data published by Maplecroft, the country was ranked as the 4th most vulnerable in the world in 2010 and it still ranks as the 16th most vulnerable in data published in 2012 (Regmi and Bhandari, 2012). Nepal's vulnerability to CC is directly associated with accelerated global warming (Practical Action, 2010), higher sensitivity to climatic factors and low adaptive capacity to deal with anticipated climate change impacts (Ministry of Environment, 2010). The large population in Nepal is already facing problems due to climate extremes, such as acute shortages of water, food and basic services (Nepal Climate Vulnerability Study Team, 2009). CC has negatively impacted the food security and rural economy by ruining crops, agricultural yields and natural resources (Bhandari, 2008). In the last two years over 1.9 million people have been adversely affected by CC and particularly climate extremes, such as flooding, drought and landslides (Nepal Climate Vulnerability Study Team, 2009; Ministry of Environment, 2010).

Governance of CC adaptation is crucial in order to ensure that financial resources and technology can be made available to the poor and most vulnerable populations. Governance issues have been a challenge to many of the least developed countries (LDCs) in effectively managing their resources. There are studies carried out by different scholars and agencies that highlight poor governance, lack of policy coordination and lack of synergy among sectors as obstacles to the mainstreaming and institutionalisation of the CC agenda into practice (Institute for Development Studies, 2007; Ayers and Huq, 2009; OECD, 2009; Ayers *et al.*, 2010; Oxfam, 2011). Despite the progress made in CC adaptation in a country like Nepal, a clear governance roadmap is needed to effectively put adaptation into practice.

This chapter outlines the major factors that trigger or hinder the processes of promoting functional CC governance in Nepal. It also outlines opportunities for better coordination and synergy among major stakeholders, including the lowest and most effective accountable units of governance in Nepal. Its aim is to examine how different factors affect the climate-governance structure and process in Nepal and how to overcome some of the governance barriers.

2. Background and literature review

The government of Nepal is a signatory to the United Nations (UN) Convention on Climate Change (UNFCCC, 1992, 2005). Nepal is also actively taking part in international and national negotiations, debates and discussions on CC issues. CC adaptation and low-carbon development are now the priorities of the government in fulfilling its commitment to the UNFCCC. The government has initiated national plans and activities to reflect the convention and associated agreements. It has established a Climate Change Council under the chairmanship of the prime minister, and a Climate Change Management Division within the Ministry of Science, Technology and the Environment (MoSTE). However, there is still a lack of clarity, within the national context, on the institutional and financial structures, mechanisms of functional coordination and capacity regarding CC (Regmi and Bhandari, 2012).

Political instability, poor bureaucratic commitment and lack of capacity and knowledge relating to CC are some factors that resulted in the governance dilemma in Nepal. Specifically, the government and different groups of Nepali stakeholders are facing challenges related to the management of resources and institutional processes in response to CC. These challenges have created obstacles in expediting the implementation of CC programmes. Furthermore, the government is facing the challenge of managing concurrent, what might be seen as, competing and/or overlapping initiatives for CC. The current national initiatives, such as the National Adaptation

Programme of Action (NAPA), the Local Adaptation Plan of Action (LAPA) and the Special Programme on Climate Resilience (SPCR) processes, are some examples.

On the other hand, some ministries, such as the Ministry of Forest and Soil Conservation and a few intergovernmental agencies, are implementing Reducing Emissions from Deforestation and Degradation (REDD+) and Clean Development Mechanism programmes without synergy and coordination with each other. This implies that there is a need now for coherent governance architecture to manage the integration of efforts and funding, and to coordinate across different elements of national effort. There is also a need for a strategy and framework on CC to streamline sporadic, overlapping and unregulated climate initiatives in Nepal.

There are research and studies carried out which demonstrate a mismatch in addressing CC responses across various tiers of governance, particularly in countries like Nepal (Institute for Development Studies, 2007; Sietz *et al.*, 2008; Ayers and Huq, 2009; OECD, 2009; Ayers *et al.*, 2010; Ghimire, 2011; Helvitas, 2011; Oxfam, 2011; Sharma. 2011). Similarly, these studies have outlined the need for governance mechanisms to facilitate, drive and navigate adaptation and mitigation actions. However, there are variations and gaps in identifying appropriate governance structures and functional mechanisms to address CC issues to particular nations of political and economic status. This strongly indicates the need to explore the governance issues further and to find ways to address them strategically in the specific national context.

3. Research objectives

This study aims to explore and analyse CC governance in Nepal in relation to CC adaptation and low-carbon development. The central argument of this chapter is that the governance structure of Nepal is not well prepared to make full use of donations/loans for CC adaptation and mitigation because nationally the demand for a CC agenda is not yet there and it is further constrained by weak governance. The chapter investigates the dynamics of policy development, fund flow and resource management with respect to CC adaptation and mitigation efforts in Nepal in order to identify critical gaps and barriers to CC governance and opportunities for the future CC regime.

4. Research methods

4.1 Methodological framework of the study

The study involved systematic review and analysis of policies, structures, mechanisms and practices across levels of the governance in relation to CC mitigation and adaptation in Nepal. The cross-sectional review and analysis provide inter-relationships and functional properties of different government bodies and stakeholders for particular issues. These cross-sections were stratified into national, district and local strata for analysing within and between relationships, functions, capacities and perceptions of the respondents.

Policy and programme documents of the government, relevant agencies and organisations regarding CC were reviewed to analyse policy and governance of CC. Similarly, a total of 46 respondents, including policy-makers and practitioners, were interviewed to map their perceptions to different issues. Among them, 29 respondents were from practitioners (working with the community, government and non-governmental organisations (NGOs) involved in climate-related projects and programmes), and 17 represented policy-makers (government officials and policy-makers) and donor communities. The respondents among the policy-makers and government officials were selected purposively based on their engagement and responsibilities within the subject of study. Similarly, these respondents were selected from a list of professionals working in the field of CC in Nepal.

The organisations were selected by using the snowball sampling technique. The final list of professionals was then stratified into three groups based on their engagement at the community level, I/NGOs and government agencies. The number of respondents was selected in a ratio of 35 per cent to the original list. However, in the case of policy-makers it was decided to include at least one respondent from each of the concerned ministries and departments where the respondents were the responsible officials for CC initiatives in the respective ministries/departments. Besides policy-makers and practitioners, 128 respondents were from communities of the Dhungegadi and Bangesaal village development committees of the Pyuthan district. Selected communities were consulted in the research process to map their perceptions regarding the impact of CC adaptation initiatives. Focused group discussions with women's and men's groups, different livelihood groups and interest groups were carried out based on a pretested checklist of questions. The information derived from the interviews and field study is part of the PhD thesis of the corresponding author and therefore serves as the output of the study.

4.2 Theoretical framework of the study

CC governance is described as that which is closely related to state and public systems and the behaviour of government, the private sector, as well as civil society and NGOs. 'It is described as a broad range of options of coordination concerning climate change adaptation and mitigation' (Fröhlich and Knieling, 2013, p. 9). CC governance in this chapter refers to the institutions, policies, plans, mechanisms (institutional, financial, coordination, information sharing, etc.) and measures adapted by the government and other stakeholders to promote a CC agenda (Meadowcroft, 2009).

The research uses a political economy analysis framework in analysing the governance of CC adaptation in Nepal. The framework proposed by Tanner and Allouche (2011) is used in this research. This framework proposes a conceptual and methodological approach for analysing the political economy of CC. It is based on the premise that 'it is necessary to understand the impact of regional and global drivers on domestic change processes given the increasingly interdependent nature of the current global governance system' (Tanner and Allouche, 2011, pp. 2–4). The framework focuses on the CC policy processes and outcomes in terms of ideas, power and resources.

5. Findings and analysis

5.1 The context and the initiatives in addressing climate change

Nepal, being highly vulnerable to climate variability and change, needs to have policies and strategies to respond appropriately to CC impacts (Ministry of Environment, 2010). The risk and vulnerability mapping carried out in Nepal shows that the majority of sectors and regions are already impacted by CC. Similarly, the projection shows that CC impacts will be severe in the future (Nepal Climate Vulnerability Study Team, 2009; Practical Action, 2009, 2010; Ministry of Environment, 2010).

The decrease in agriculture productivity, increased frequency of disasters, water scarcity and disease epidemics have dominated disaster events in the past (UN Development Programme (UNDP), 2009). Studies in Nepal reveal that over 1.9 million people have been severely affected by similar events (Nepal Climate Vulnerability Study Team, 2009; UNDP, 2009; Regmi and Bhandari, 2012). The strong linkages between poverty and CC are already evident in the Nepali context (Gentle and Maraseni, 2012).

The climate of Nepal is diverse and is influenced by a complex mountainous topography and landscape. The country's diverse landscape nurtures several unique and important microclimatic areas. The CC study shows that temperatures are increasing and rainfall is more variable in Nepal. Similarly, the alteration in temperature and precipitation observed in the last 30 years shows that the change is uneven and very varied (Practical Action, 2009). National reports also show that the rate of warming in the Himalayas is greater than the global average, confirming that the Himalayas are among the regions most vulnerable to CC (Practical Action, 2009; Shrestha, Gautam and Bawa, 2012). Similarly, Nepal's major social and economic sectors are likely to be impacted by climate change (Ministry of the Environment, 2010). The impact has already been observed in agriculture and weather-related hazards, such as drought, extreme rainfall and forest fires (Alam and Regmi, 2004; Bhandari, 2008; Malla, 2008).

Nepal has made good progress with its CC agenda. A number of policy measures have been endorsed by the government of Nepal to support CC initiatives (Ministry of the Environment, 2010, 2011a, 2011d; Ministry of

Science, Technology and Environment, 2012a, 2012b). CC policy has been formulated to guide Nepal to respond effectively in adaptation and to adopt a low-carbon development pathway (Ministry of the Environment, 2011b). Similarly, the government has also formulated the NAPA. It has recently endorsed the LAPA in order to facilitate the implementation of adaptation priorities (Ministry of the Environment, 2011c).

In addition to these policies, institutional arrangements have been put forward by the government to facilitate CC. The establishment of the Climate Change Council, the Multi-stakeholder Climate Change Initiatives Coordination Committee and the Climate Change Division are some examples of institutional efforts towards mainstreaming CC. The analysis shows that the government has attempted to respond to CC in a number of ways and with support from development partners, such as the Asian Development Bank (ADB), the World Bank (WB), the UK Department for International Development (DFID) and the European Union. However, most of these attempts have been limited to policy design at the central/national level and lack local-level initiatives and actions on ground.

5.2 Issues surrounding climate change governance

5.2.1 International policy shaping national climate-change policies and programmes

This research looked into the historical perspective of policy-making in order to draw inferences with the governance of policy development in Nepal. The findings from the historical trend analysis of CC evolution in the country show some kind of deliberate influence of international policies and actions. For example, most of the treaties on biodiversity, CC, food security and MDGs have been prepared based on international influences. Unlike the forestry sector policy advancement in Nepal where the policy-making has been demand driven, CC policy and planning have been greatly influenced by the international process, such as the UNFCCC negotiations and aid dynamics.

The physical risks had little impact on CC discussions in the past. Nepal has meteorological records from across the country only since 1976, which is not sufficient to draw conclusions on the long-term climate trend. Available data are also questioned for their reliability. Therefore the issues are largely based on the information from global and regional data and the observed climate-related impacts on the ground. The impact of CC has been noted since the 1980s, when the rural areas of Nepal experienced damage due to glacier lake outburst floods (International Centre for Integrated Mountain Development, 2011). These events were linked to natural disasters and remained unheard of until the late 1990s. Most of the policy documents in the 1990s and early 2000 focused on emissions reduction and addressing risk in line with the international context.

The policy-making process, institutional structure and financial resources are all related to how the international negotiation and policy-making process is shaped. The government of Nepal formulated a pro-environment plan during the 6th Five Year Plan (1980–1985) as a response to meet the requirements of the UN Conference on the Human Environment in 1972 (National Planning Commission, 1980). Since the Earth summit, Nepal developed a number of environmental frameworks and policies along with decentralisation policies and programmes. The policy-making process, therefore, could be linked and attributed to how the international agenda is shaped.

Another context of policy governance is the influence of international negotiation and policy process at the national level. In between 1990 and 1996, the convention on CC was very optimistic regarding global action to voluntarily reduce emissions. It proved the international CC policy-makers wrong so they adapted the Kyoto Protocol to make a legally binding agreement for developed and industrialised nations to comply with emissions reduction. The international policy-makers were confident that the emissions-reduction target would be achieved. However, the complex economic and political dynamics between the nations complicated the negotiation and compliance processes. The Kyoto Protocol achieved less as major economies blocked the negotiations. In this period, Nepal's government was passive in taking strategic decisions as there was no international pressure and country-specific agenda stimulating it to take action. There was also less public awareness, low donor appetite and less government priority on dealing with CC. This was reflected in a lack of any action plans or activities during the 1990s and the beginning of 2000.

There was some active progress within countries, including Nepal, after the introduction of the adaptation regime into the global climate negotiation in 2007. Many stakeholders realised that adaptation came into existence because of the fact that mitigation was complicated and not achievable to the required level. There were also reports and evidence from some LDCs showing the magnitude and impact of CC. The impact also jeopardised the UN's Millennium Development Goals (MDGs). This might be the reason why negotiators of the LDCs realized the significance of adaptation agenda. It could also be because of the fact that the group involved in adaptation grew larger with support from some environmental organisations and donors (Burton, 2009).

The findings from the interaction with practitioners and policy-makers in this research project also revealed that the CC agenda is externally driven and shaped by international negotiations. One of the respondents among the policy-makers said: 'as climate change is imported, it is difficult to get national level ownership of the issue' (respondent number 5). Similarly, half of the practitioners interviewed (19) also felt that country ownership of the agenda is crucial to get both public as well as political support. These responses imply that, nationally, there is a dominant perception that

CC is a global problem and that the developed world is responsible for solving it.

The analysis, discussed in the above paragraphs, implies that the CC agenda evolution in Nepal is more influenced by international demand than the country's need. It shows that international agreement and particularly the commitment of Nepal shaped the inclusion of the CC agenda within national debates and discussions. It shows that the compliance agenda actually failed to generate momentum on CC with the national policies and system.

5.2.2 Influence of development aid on shaping the climate-change agenda

Development aid has influenced the development agenda of the developing countries. Development assistance, aid and CC now form the main discussion issues among policy-makers at both the international and national levels. The findings show that development aid in Nepal is largely dependent on donor countries. Similarly, government records show that aid comprises 60 per cent of the total development budget (National Planning Commission, 2012). The way in which climate action has advanced in Nepal has a strong connection to development aid and policies and programmes. Until 2009, only a few I/NGOs were implementing activities or advocating CC action in Nepal. There were only five major donors in the mainstream CC discussion (the United Kingdom's Department for International Department, the Danish Aid Agency, the ADB, the WB and the UNDP) until 2009. This number grew and reached 15 by 2009–2010 (Oxfam, 2010). The donor compact, signed by the development partners and the government of Nepal in 2009, was the outcome of the commitment that might be the factor triggering many donors to be interested in CC issues in the country.

There were positive as well as negative views and perceptions of policy-makers and practitioners in terms of donor engagement. For reasons that aren't clear, the donor communities have embedded CC within their portfolios. Many scholars criticise this move as loss to the bilateral funding basket (Bouwer and Aerts, 2006; Klein *et al.*, 2007; Bapna and McGray, 2008; Ayers and Huq, 2009). The labelling of CC to the conservation and development funds also had major implications for the sort of funding available for other key priorities, such as food security, poverty reduction and biodiversity conservation. For example, the United States Agency for International Development (USAID) has launched a biodiversity and CC project recently, the Hariyo Ban Program, which moved away from its original focus of biodiversity and livelihood to biodiversity and CC. Similarly, the UK, Finnish and Swiss donors have also used their forestry project resources for CC activities. The Multi Stakeholder Forestry Programme, launched by these donors in 2011, has included CC as one of the key strategic focuses. The majority of the interviewed participants from donor

Table 14.1 Major donor-supported climate-change policy and programme development

Year	Policies/programme	Supported by
2004	First National Communication Report to Conference of Parties of the UNFCCC	UNFCCC through UNEP
2009	National Strategy for Disaster Risk Management	European Commission, humanitarian aid, UNDP
2009	NAPA	LDCF/GEF, UNDP, DFID and DANIDA
2010	Strategic Programme for Climate Resilience	ADB and WB
2011	Climate-Change Policy	World Wildlife Fund
2011	LAPA Framework	DFID
2011	Mountain Initiatives	multiple donors
2011	Second National Communication Report to Conference of Parties of the UNFCCC	UNFCCC through UNEP
2012	Climate-Resilient Framework	ADB
2009–2012	various position papers for the UNFCCC meeting, including Rio 20+	DANIDA, DFID, WB, ADB

Source: Outcome of desk review carried out by the authors.

communities (four out of five interviewees) perceived this as integration and mainstreaming efforts of the donors, while civil society representatives (all three interviewed) viewed these shifts as attempts to override other development priorities.

The debate about aid influence in the development of policies and a strategic programme is a reality in Nepal. The analysis of the major policies and landmarks related to CC and the environment reveals the donor investment and support in policy drafting and preparation. The major policies and programmes in Nepal were formulated with support from donors, such as environment policy, CC policy, the Agriculture Perspective Plan, the Master Plan for the Forestry Sector and the National Strategy for Disaster Risk Management. Table 14.1 shows that the major CC-related policy and programme were prepared with donor support. This implies that the resource investment trend in Nepal is externally influenced through aid and the international agenda in the policy-making process. Similar findings were reported by other authors (Gautam and Pokharel, 2011; Ghimire, 2011).

The above analysis shows that there is a deliberate influence of development and bilateral aid on the development of the CC agenda in Nepal. Although the international financing emphasised the equity issues, it appeared that the agenda was shaped according to donors' and development partners' interests. The government of Nepal, due to budget deficits, largely depends on aid and external resources, so it is not influential in shaping CC governance.

5.2.3   Financial dilemma

The financial mechanism and process have a major impact on CC governance of the country. In Nepal the major development partners, such as the ADB and WB, have dominated aid throughout the decade (Ministry of Finance, 2010). The aid effectiveness review shows that the bilateral aid from the DFID and India has grown significantly, while the level of aid from Japan has stagnated, if not decreased. The total foreign aid remains at an average of 20 per cent of the national budget with the highest being 24 per cent in 2000/2001 and the lowest 18 per cent in 2007/2008 (Ministry of Finance, 2010). The report from the Ministry of Finance shows that foreign aid rose by 77.9 per cent in the fiscal year 2009/2010 compared with the fiscal year 2008/2009 (Ministry of Finance, 2010). However, this significant increase has not raised the proportion of foreign aid in the budget. This clearly shows that there was no substantial increase in the overall bilateral/multilateral funding despite the recent international claim by developed countries in the Conference of Parties' 18th meeting in Doha about its increase in climate financing for the LDCs.

The Climate Public Expenditure Review carried out by the UNDP and the UN Environment Programme (UNEP) shows that there are 13 programmes, with a total cost of US$326 million waiting for funding from various sources (Bird, 2011). The proportion of government versus donor funding varies within the five-year period. Government funding for CC was 54.2 per cent in 2007/2008 but it decreased to 44.1 per cent in 2011/2012. In contrast the proportion of donor funds increased from 20.7 per cent in 2007/2008 to 40.4 per cent in 2011/2012. The increase in the amount of donor funding in 2011/2012 is due to the existence of CC policy and a NAPA, which made donors the basis of investing their resources in CC. Of this amount, approximately US$225 million is in the form of grants and about US$101 million is loans. This ratio of grant to loan is approximately 30:70 per cent (Bird, 2011). Loans are a decreasing trend. This could have serious financial and moral implications in meeting both development and CC goals. There might also be chances of diversion of committed development budget to count for CC support.

The appearance of multidonor trust funds generated some controversy in climate financing. There are issues in terms of loan acceptance in CC adaptation. The SPCR provided US$86 million in 2011 to Nepal to implement a climate-resilience programme, out of which US$36 million was loaned (Regmi and Bhandari, 2012). This has drawn a lot of criticism of the government and donors. Civil society and the media have strongly advocated against loans in CC adaptation. They have raised concerns over the violation of the UNFCCC agreement on the liability of developed countries to support poor countries. The donors, particularly the WB and the ADB, were interested in bringing private-sector investment in the SPCR through loan arrangements.

Despite public protest, the government decided to accept the loan. One of the interviewees from the Ministry of the Environment expressed his dissatisfaction over the process, stating that

> our government does not have any clear strategy on loan and grant. We were forced to accept the condition as the multilateral banks are major bilateral donors in Nepal. Our officials also lack clear understanding of our role in international negotiation for e.g. on whether we should accept loan for climate change adaptation or not.
>
> (Interview with an official from the Ministry of the Environment, 2012)

The civil society representatives also shared similar views about loan acceptance, stating:

> government lack clear vision on managing climate financing in Nepal. Accepting loan on adaptation contradicts with the international agreement of liabilities on adaptation.
>
> (interview with Civil Society Groups, 2012)

This clearly demonstrates the weak position of the government and the donors' indirect influence on funding. This case was also debated as a climate justice issue within national and international communities (Ghimire, 2011).

The mistrust regarding climate financing at the international level has also affected national-level debates, discussions and practical initiatives on the ground. There is evidence showing that the donor agencies have not met their Overseas Development Assistance commitment of 0.7 per cent of gross national income to developing countries (Gupta and Van der Grijp, 2010). There is a lack of clarity regarding what mode of financing is feasible and effective in terms of CC mitigation and adaptation. The existing financial arrangements within the international regime cannot alone serve as an adequate basis for achieving the much-needed assistance for adaptation in the most vulnerable countries so that the bilateral mode of supporting adaptation can be meaningful (Ayers and Huq, 2009). However, there are diverse views and interpretations among academic scholars and government in different countries regarding the financial architecture and use of bilateral aid.

The analysis shows that CC financing is fragmented and not institutionalised in Nepal. The report published by Oxfam (2011) stated that climate financing there is fragmented and disconnected. There were also trends among donors to use their bilateral commitment for development of CC action. Many donors strategically justified their development and resource management projects as relating to CC. For example, the list of donor support for CC appeared high in the donor funding list in contrast to only half a dozen projects with a real CC focus. Similarly, there were issues around budgeting trends. A significant sum of technical assistance provided by donors – US$13 million per year – for climate support was not budgeted or accounted for through government systems (i.e. 'off budget'). The major donors, such as USAID, the UK and Japan, provided only 30 per cent of their bilateral aid on-budget (Bird, 2011).

The analysis of the financial flow and mechanism mentioned above revealed that it has major implications for CC governance, particularly on the issues of transparency and accountability of fund management. These issues are often debated informally within and between governments, civil society and donor agencies. Despite the concern almost everywhere, the situation has not improved. The parallel governance structure created through bilateral and beyond the national budget provisions has created a dilemma for effective CC governance. There is a need for the intensive engagement of stakeholders to overcome the barriers that limit the effectiveness and efficiency of harmonising policies and funding to CC adaptation and mitigation. The process should open the door to discussions about collective agreement to a governance model to address the issues associated with harmonising financial flow.

5.2.4 Policy and institutional dilemma around agencies and their role

Governance is an important factor in attracting and utilising international assistance where national capacity is not sufficient to meet domestic requirements. Institutional and governance barriers also need to be tackled in terms of climate financing as well as in other development actions. Similarly, the role of government, the private sector and civil society becomes important to ensure effective governance within the country. This section analyses the key policy and institutional barriers to CC governance in Nepal.

This research project shows that there are policy dilemmas in CC. The only policy document that talks about CC adaptation is the CC policy. This was endorsed by the cabinet in 2011. The process adopted was criticised by many civil society and community groups as it failed to consult wider stakeholders at the interest of international agency-dominating authorities in the ministry (Helvitas, 2011). The interviews carried out with the communities revealed that more than 99 per cent (out of 128) of the participants were not aware of the existence of the CC policy. Similarly, the policy focus was diluted by the influence of the international agenda, such as regarding low-carbon development. Despite the country's need to address vulnerability, the CC policy has not given enough emphasis to adaptation. Similarly, no progress has been made so far in implementing the strategies and action plan laid out by the CC polices.

There are issues around institutional governance related to managing CC in Nepal. The MoSTE, although recognised as the nodal agency for coordination, is yet to be operationally recognised by other line ministries. The discussion with policy-makers and the review of the documents, particularly the programmes and plans in different sectoral ministries and departments, indicated that there is a lack of trust, and clarity on roles and responsibilities. There is also a coordination gap between government agencies. The MoSTE does not have institutional access to district and community levels, and there are limited human resources to coordinate CC issues. Therefore the ministry has to rely on other frontline agencies to plan, implement and monitor environmental programmes at the local, district and national levels. It is anticipated that the ministry should facilitate the building capacity of other ministries and departments in integrating CC adaptation and mitigation into their sectoral strategies, programmes and actions. However, the paper of Regmi and Bhandari (2012) argue that MoSTE is not properly coordinating CC and could not institutionalise the issues with other ministries and departments. Other ministries see this gap as a barrier to the coordination of cross-cutting issues, such as CC (Regmi and Bhandari, 2012).

The study by Regmi and Bhandari (2012) found that there were also issues around institutional governance due to a weak monitoring system and law enforcement. Institutional governance was further damaged due to strong political interferences and corruption cases. This finding is also supported by Khadka *et al.* (2012, p.13), who argued that 'political instability for the last two decades and current stalemate over agreeing a new Constitution has inhibited government agencies to focus on mainstreaming agenda'. The respondents from district-level agencies, and community and civil society practitioners, outlined that there has been a lack of coordination and communication between ministries, departments and district-level agencies. This lack of coordination, according to the respondents, is a barrier in effectively mainstreaming CC into relevant sectors. Similarly, the respondents felt that the gaps include timely budget flow, sufficient budget allocation, technical inputs and the designation of the right person in the right place.

There are also issues around transparency and accountability within the government system. The interviews with policy-makers in this research also revealed that Nepal has to do more in order to establish a transparent and accountable financial system to improve its transparency index. There are also studies which show that many least developed and developing countries, mostly in Africa and Asia, are facing a similar problem (Sietz *et al.*, 2008; Lasco *et al.*, 2009; Senaratne, Perera and Wickramasinghe, 2009).

The multifaceted nature of CC requires strong collaboration and synergy among government and non-government actors, and more importantly collective action to deal with the issue. The findings show that the government of Nepal has recognised the role of multistakeholders in CC design and delivery. The framework on LAPA has emphasised the role of local government, civil society, the private sector and communities in implementing CC adaptation. However, in practice it is not observed as there are examples where government and non-government sectors are often competing for resources and duplicating the works. There are different programmes related to disaster risk reduction and CC adaptation and mitigation (CDM on biogas and REDD+) operating in different parts of Nepal. But these projects are implemented in isolation and often lack linkages and synergy with each other (Sharma, 2011).

The findings in the above paragraphs highlight key barriers related to institutional structures, coordination mechanisms and the conflicting roles of various groups of stakeholders in the case of Nepal. It clearly shows the mismatches between policies and institutional structures, creating obstacles in ensuring good implementation of CC priorities in Nepal.

5.2.5 *Knowledge and capacity on assessment and addressing climate change*

The nature of CC demands innovative knowledge and capacity to respond to the issue. Nepal is facing a brain-drain problem and losing its trained and skilled human resources due to political transition and instability among others. Similarly, there are limited educational opportunities and research centres within the country to initiate CC research and studies. The lack of information in Nepal forces it to rely on regional circulation models and research carried out by regional research institutions and international organisations. As the country harbours unique microclimatic areas, it is unlikely that the down-scaled models are able to represent the country's CC reality. Similarly, most of the experts on CC come from outside and skill transfer is lacking. The government sectors also lack expert, skilled and trained human resources in CC. This has created a knowledge vacuum and ultimately impacted the CC governance in Nepal.

Similarly, Nepal is considered to be the white spot in the Intergovernmental Panel on Climate Change's report in terms of the lack of a historical and reliable database on weather and climate. At the national level there is also a problem in terms of accessing meteorological database and CC information. Nepal has no more than 180 functional meteorological stations which will provide information about temperature and rainfall trends since 1976 only. The complex nature of CC risk and vulnerability cannot be explained solely by temperature and rainfall data. Socioeconomic and ecological diversity and dynamics play an important role in shaping the risk and vulnerability context. This implies that downscaling of information is important in order to generate reliable information for informed decision-making.

There are also issues around downscaling CC information and making the best of it for policy- and development-based decision-making. The analysis shows that a lack of information at the sectoral level was a constraint in terms of convincing the sectoral and development ministries to take firm action in responding to CC. Furthermore, there are issues of misinterpretation. Regmi and Bhandari (2012) argue that 'while observed changes on the ground were interpreted and attributed forcefully to CC alone, future predictions were made based on the institutions that misguided the science such as by forecasting the life of Himalayan permafrost to be ended by 2035' (p. 49). Such predictions are likely to decrease the trust on scientific climate forecasts, inviting the story of 'crying wolf' into reality. This implies that both the source and the reliability of information and knowledge influences policy-decision making, governance structure and action on CC.

6. Discussion

The above analysis presents complex scenarios of CC governance with reference to Nepal. These governance challenges observed in Nepal are similar to those in other developing nations. Many regions in the developing world where current and projected poverty levels are high, coincidentally, are those regions where CC effects are projected to hit earlier and worse than elsewhere. Adaptation discourse is far behind many development debates, potentially recreating some mistakes, such as projectised, donor-driven and uncoordinated assistance and top-down planning. Although the conventional development models have created some basics of governance and decision-making, they have not been effective in many low-income countries.

The international dynamics of fragmenting CC has serious implications for the governance of CC at the national level. This implies that we need to strategically reform our institutions and governance mechanisms. This also demands a shift from building resilience towards transformational change. This shift, as described by Oxfam (2010), is needed to move communities from being victims of CC to actively pursuing opportunities and allaying the negative consequences of CC. This shift will bring desirable change in both society and households, as evident in the case of Bangladesh where communities have taken the leadership in addressing CC issues, to sustain livelihoods in a changing climate (Regmi and Bhandari, 2012). Furthermore, adaptation is a continuous process and demands flexibility, innovation and robustness in the system and institutions to systematically organise and respond to adverse impacts.

Similarly, the discussions of adaptation governance, at the national level, stress the need for innovating knowledge and a radical shift in the mindset of policy-makers and practitioners. Armitage and Plummer (2010) and Regmi and Bhandari (2012) argue that in a country where the ecological, social and economic conditions are untenable, there will be limits to adaptation and a need for more fundamental shifts in strategy that require new ideas and practices. Adaptation to CC presents a complex methodological challenge due to the traits of adaptation challenges, such as uncertainty, complexity, irreversibility and urgency (Claycomb, 2009). This makes adaptation a difficult problem that requires innovative approaches and perspectives of analysis (Brown, 2011).

Recent academic discussion has focused on the need for society in the LDCs to radically transform their institutions, technologies, systems and practices to adjust to CC. The transformational change is not just required at the community level but also applies to the institutional and policy-making process and its governance and implementation. Adger (2003) revealed that the adaptation process involves the diverse nature of stakeholders and their relationships with the institutions where they reside and the resource base that they depend on. This understanding of the relationship is needed to foster collective action in adaptation in the context of LDCs like Nepal.

The findings also show that there is huge scope for bringing different sectors and communities together under one umbrella to manage CC governance in Nepal. Clearly the initiatives can be complementary if they are prioritised and managed coherently. The governance of all CC activities, particularly the major initiatives, should mirror the governance of climate action and financing more broadly, building and strengthening what already exists.

Furthermore, the intensity of negotiation in international conferences and subsequent meetings is already putting pressure on national government and stakeholders to be more strategic and responsive to the needs of millions of vulnerable households. The responses are delayed and frustration is growing within communities and practitioners who want to see nothing but action on CC. The national government and bilateral donors now need to work together to mobilise in-country resources to provide immediate support to vulnerable household and communities.

7. Limitations

This study focuses on CC governance and has not been able to diagnose the broader governance and development paradigm. Due to the limitations of national-level literature, the analysis is mostly based on case studies and the researchers' knowledge and experiences to interpret the policies and practices. Similarly, this study was not able to track the dynamic nature of policy and governance function due to its short duration. However, the authors believe that this chapter provides a good analysis of CC governance and strong reference to future studies in exploring elements of good governance in CC adaptation.

8. Conclusion

The findings show that the governance structure of Nepal is not well pre-pared to make full use of international support for CC adaptation and mitigation. This is because, nationally, the demand for a CC agenda is not yet there and it is further constrained by weak governance. The governance of CC in Nepal is influenced by international agendas, donor and devel-opment agencies' interest, financial dilemmas, and a lack of information and knowledge about CC. The following section recommends strategies to address governance issues.

8.1 Enhancing climate change governance: Building on and shift to good practice

Societies and communities in developing countries do not have sufficient resources and capacity to transform quickly. It is evident that facilitating the gradual change, with support from the state and international community, can be an important entry point. The national and local context demands an expansion of existing knowledge, innovations and best practices to facilitate the exchange of information and knowledge so that it can be understood widely. Flexibility and innovativeness in the process of facilitating adapta-tion is important. There are already some good practices and innovative institutional models tested in the development sector in Nepal that can be of help to CC adaptation governance. The lesson from community-based natu-ral resource management in Nepal is important for policy-makers to learn in understanding the multifaceted nature of resource governance and the role of various actors and agencies in response measures (Regmi and Bhandari, 2012).

8.2 Synergy between local, national and international negotiation

The analysis shows that Nepal is preparing itself to address CC issues along with other LDCs. However, there is a lack of clarity and a dilemma around institutional and financial governance structure to drive the agenda forward. There are specific governance issues around fund flow and management, capacity and human resources, and information and knowledge gaps. Simi-larly, the findings show that there has not been serious discussion around climate governance in Nepal. The resources are starting to flow and the bilateral agencies and government are confused about agreeing to a com-mon mechanism for fund flow. As a consequence, the investment is ad hoc and fragmented. This suggests that it is necessary to devise a strategic roadmap and architecture on climate financing and governance in Nepal, supported by strong policy and multistakeholder commitment. The funding has to be decentralised and easily accessible to vulnerable household and communities.

8.3 Building synergy between institutions and sectors

It is revealed by the study that designing a new mechanism and institutional structure in CC governance may demand strategic thinking and input. In order to streamline the efforts now and to manage the resources and expectations, the government of Nepal can start to work on a phase-wise approach to ensure an effective governance setup. What is needed at this stage is a mechanism jointly agreed among government, development agencies and civil society to coordinate the initiatives and create a multistakeholder forum or platform to make strategic decisions about fund management and implementation. There could be a temporary management structure and fund-flow mechanism, agreed among stakeholders, in order to avoid delays in implementation.

The necessary process is as follows:

- Step one (preparation stage) – to ensure that major stakeholders in Nepal realise the need for harmonisation and synergy among CC initiatives.
- Step two (transitional stage) – setting up a transition structure to allow gradual harmonisation which focuses on awareness raising, knowledge generation and capacity-building.
- Step three (harmonisation stage) – programme and projects are streamlined through the agreed structure.
- Step four (implementation stage) – where a well-coordinated and mutually owned structure and mechanism are functional, and communities benefit from the well-governed structure.

Abbreviations

ADB	Asian Development Bank
CC	climate change
DFID	Department for International Development
DANIDA	Danish International Development Agency
GEF	Global Environment Facility
I/NGO	international/non-government organisation
ICIMOD	International Centre for Integrated Mountain Development
LAPA	Local Adaptation Plan of Action
LDC	Least Developed Countries
LDCF	Least Developed Countries Fund
MoSTE	Ministry of Science Technology and the Environment
NAPA	National Adaptation Programme of Action
REDD+	Reducing Emissions from deforestation and degradation
SPCR	Special Programme on Climate Resilience
UN	United Nations
UNFCC	UN Framework Convention on Climate Change

UNDP   UN Development Programme
UNEP   UN Environment Programme
USAID   United States Agency for International Development
WB   World Bank

References

W.N. Adger (2003) 'Social Capital, Collective Action, and Adaptation to Climate Change', *Economic Geography*, 79, 387–404.

M. Alam and B.R. Regmi (2004) *Adverse Impacts of Climate Change on Development of Nepal: Integrating Adaptation into Policies and Activities* (Dhaka: Bangladesh Centre for Advanced Studies).

D. Armitage and R. Plummer (2010) 'Adapting and Transforming: Governance for Navigating Change', in D. Armitage and R. Plummer (eds.) *Adaptive Capacity and Environmental Governance* (Berlin Heidelber: Springer series on environment management).

J. Ayers, M. Alam and S. Huq (2010) 'Global Adaptation Governance Beyond 2012: Developing Country Perspectives', in F. Biermann, P. Pattberg and F. Zelli (eds.) *Global Climate Governance Post 2012: Architecture, Agency and Adaptation* (London: Cambridge University Press).

J. Ayers and S. Huq (2009) 'Supporting Adaptation to Climate Change: What Role for Official Development Assistance?' *Development Policy Review*, 27, 675–692.

M. Bapna and H. McGray (2008) *Financing Adaptation: Opportunities for Innovation and Experimentation* (Washington D.C.: The World Resources Institute).

D. Bhandari (2008) 'Living with Uncertainty: Climate Change and Disasters', in B. Regmi (ed.) *Melting Glaciers* (Pokhara: Local Initiatives for Biodiversity, Research and Development (LIBIRD).

N. Bird (2011) *The Future for Climate Finance in Nepal* (London: Overseas Development Institute).

L.M. Bouwer and J.C.J.H Aerts (2006) 'Financing Climate Change Adaptation', *Disasters – Special issues on Cliimate Change and Disasters*, 30, 49–63.

K. Brown (2011) 'Sustainable Adaptation: An Oxymoron?' *Climate and Development*, 3, 21–31.

I. Burton (2009) 'Deconstructing Adaptation and Reconstructing', in L. Schipper and I. Burton (eds). *The Earthscan Reader on Adaptation to Climate Change* (London: Earthscan).

Civil Society Groups (2012), *Interview with Members of Civil Society Groups* (Kathmandu-Nepal).

P. Claycomb (2009) 'SAS 2: A Guide to Collaborative Inquiry and Social Engagement', *Development in Practice*, 19, 1091–1093.

J. Fröhlich and J. Knieling (2013) 'Conceptualising Climate Change Governance', in J. Knieling and W.L. Filho, Springer (eds) *Climate Change Governance* (Berlin Heidelberg: Springer).

M. Gautam and B. Pokharel (2011) *Foreign Aid and Public Policy Process in Nepal: A Case of Forestry and Local Governance* (Kathmandu: Southasia Institute of Advance Studies).

P. Gentle and T.N. Maraseni (2012) 'Climate Change, Poverty and Livelihoods: Adaptation Practices by Rural Mountain Communities in Nepal', *Environmental Science and Policy*, 21, 24–34.

S. Ghimire (2011) *Climate Justice: Bottlenecks and Opportunities for Policy-Making in Nepal* (Kathmandu: Southasia Institute of Advanced Studies).

J. Gupta and N. Grijp (2010) *Mainstreaming Climate Change in Development Cooperation: Theory, Practice and Implications for the European Union* (London: Cambridge University Press).

I. Helvitas (2011) *Nepal's Climate Change Policies and Plans: Local Communities' Perspective, Environment and Climate Series 2011/1* (Kathmandu: Helvitas-Swiss Intercorporation Nepal).

S. Huq and H. Reid (2004) 'Mainstreaming Adaptation in Development', *IDS Bulletin*, 35 (3), 15–21.

International Centre for Integrated Mountain Development (2011) *Glacier Lakes and Glacier Lake Outburst Floods in Nepal* (Kathmandu: ICIMOD).

Institute of Development Studies (2007) 'Mainstreaming Climate Change Adaptation in Developing Countries', *IDS in Focus*, 2, 1–2.

R.B. Khadka, D.B. Clayton, A. Mathema, and P. Shrestha (2012) *Safeguarding the Future, Securing Shangri-La – Integrating Environment and Development in Nepal: Achievements, Challenges and Next Steps* (UK: International Institute for Environment and Development).

R. Klein, S. Eriksen, L. Naess, A. Hammill, T. Tanner and C. Robledo (2007) 'Portfolio Screening to Support the Mainstreaming of Adaptation to Climate Change into Development Assistance', *Climate Change*, 84. 23–44.

R.D. Lasco, F.B. Pulhin, P. Jaranilla-Sanchez, R.J.P. Delfino, R. Gerpacio, K. Garcia (2009) 'Mainstreaming Adaptation in Developing Countries: The Case of the Philippines', *Climate and Development*, 2, 130–146.

G. Malla (2008) 'Climate Change and Its Impact on Nepalese Agriculture', *The Journal of Agriculture and Environment*, 9, 62–71.

J. Meadowcroft (2009) *Climate Change Governance Policy Research Working Paper Series* (Washington D.C.: World Bank).

Ministry of the Environment (2010) *National Adaptation Programme of Action, Ministry of Environment* (Kathmandu: Government of Nepal).

Ministry of the Environment (2011a) *District Climate Change and Environment Management Guidelines* (Kathmandu: Government of Nepal).

Ministry of the Environment (2011b) *Climate Change Policy 2011* (Kathmandu: Ministry of the Environment).

Ministry of Environment (2011c) *National Framework on Local Adaptation Plan for Action (LAPA)* (Kathmandu: Government of Nepal).

Ministry of Environment (2011d) *Status of Climate Change in Nepal* (Kathmandu: Ministry of Environment).

Ministry of Environment (2012) *Informal Interview with Official in MoE* (Kathmandu: Ministry of Environment).

Ministry of Finance (2010) *Joint Evaluation of the Implementation of the Paris Declaration, Phase II: Nepal Country Report* (Kathmandu: Government of Nepal).

MoSTE (2012a) *Community Based Vulnerability Assessment Tools and Methodologies and Risk Mapping* (Kathmandu: Government of Nepal).

MoSTE (2012b) Submission from Nepal on Financial Mechanism of the Convention – Least Developed Country Fund. Subsidiary Body for Implementation of the United Nations Conference on Climate Change. Submitted on August 1, 2012, published on September 25, 2012, http://unfccc.int/resource/ docs/2012/sbi/eng/misc12.pdf, date accessed 3 November 2012.

Nepal Climate Vulnerability Study Team (2009) *Vulnerability Through the Eyes of Vulnerable: Climate Change Induced Uncertainties and Nepal's Development Predicaments* (Kathmandu: Institute for Social and EnvironmentalTransition-Nepal).

National Planning Commission (1980) *Sixth Five Year Plan (1981–1985)* (Kathmandu: Government of Nepal).

National Planning Commission (2012) *Tenth Five Year Plan Progress Review* (Kathmandu: Government of Nepal).

OECD (2009) *Integrating Climate Change Adaptation into Development Co-Operation Policy Guidance* (Paris: Organisation for Economic Cooperation and Development).

Oxfam (2010) *The Impact of the Global Financial Crisis on the Budgets of Low-Income Countries* (UK: Oxfam International).

Oxfam (2011) *Minding the Money: Governance of Climate Change Adaptation Finance in Nepal* (Nepal: Oxfam).

Practical Action (2009) *Temporal and Spatial Variability Climate Change over Nepal (1976–2005)* (Kathmandu: Practical Action).

Practical Action (2010) *Impacts of Climate Change: Voices of People* (Kathmandu: Practical Action).

B.R. Regmi and A. Adhikari (2007) *Climate Change and Human Development-Risk and Vulnerability in a Warming World: HDR Nepal Case Study, Occasional Paper* (Washington D.C.: United Nations Development Programme).

B.R. Regmi and D. Bhandari (2012) 'Climate Change Governance and Funding Dilemma in Nepal', *TMC Academic Journal*, 7, 1, 40–55.

A. Senaratne, N. Perera and K. Wickramasinghe (2009) *Mainstreaming Climate Change For Sustainable Development in Sri Lanka: Towards a National Agenda for Action* (Colombo: Institute of Policy Studies).

K. Sharma (2011) 'The Political Economy of Climate Change Governance in the Himalayan Region of Asia: A Case Study of Nepal', *Procedia-Social and Behavioural Sciences*, 14, 129–140.

U.B. Shrestha, S. Gautam and K.S. Bawa (2012) 'Widespread Climate Change in the Himalayas and Associated Changes in Local Ecosystems', *PLoS ONE*, 7, 367–41.

D. Sietz, M. Boschutz, R. Klein and A. Lotsch (2008) *Mainstreaming Climate Adaptation into Development Assistance in Mozambique: Institutional Barriers and Opportunities* (Washington D.C.: World Bank).

T. Tanner and J. Allouche (2011) 'Towards a New Political Economy of Climate Change and Development', *IDS Bulletin*, 42, 1–14.

UNDP (2009) *Nepal Country Report, Global Assessment of Risk* (Kathmandu: United Nations Development Programme).

UNFCCC (1992) United Nations Framework Convention on Climate Change, http://unfccc.int/essential_background/convention/status of ratification/items/2631.php, date accessed 22 September 2012.

UNFCCC (2005) United Nations Framework Convention on Climate Change, http://unfccc.int/kyoto_protocol/status_of_ratification/items/2613.php, date accessed 22 September 2012.

15
Environmental Legislation and Action in Polity, Economy and Culture for Climate Change Adaptation: A Case Study of Misamis Oriental Province, the Philippines

Isaias S. Sealza and Huong Ha

1. Introduction

Climate change (CC) caused concern when the rains no longer came with clockwork regularity, when the heat became oppressive and the effects of *El Niño* and *La Niña* became vicious. In accordance with international treaties, notably the United Nations (UN) Framework Convention on Climate Change (UNFCCC) and the Kyoto Protocol, the Philippine national government enacted landmark legislation, such as the Philippine Clean Air Act in 1999, the Solid Waste Management Act in 2000, the Biofuels Act in 2007 and the Renewable Energy Law in 2008. It formulated the Philippine Climate Change Response Framework (CCRF), which highlighted the need for the effective governance of CC response, proaction of stakeholders, lifestyle change and action within international communities (Presidential Task Force on Climate Change, 2007). It created a Climate Change Commission to coordinate, monitor and evaluate CC-related programmes and action plans of the government (Loft and Kenny, 2012). Subsequently, it launched the CC academy for education in disaster-risk reduction and management.

The twin responses to CC are mitigation and adaptation. Mitigation requires cutting the root causes of CC. Obviously it entails a global movement with the introduction of the UNFCCC and the Kyoto Protocol. Adaptation, on the other hand, is a local response to sustain life in spite of CC. Active adaptation also implies mitigation, as when a farmer adapts to CC by replacing annual staple crops with perennial fruit trees – they have to contribute to reducing the causes of CC.

While the national government can enter into international agreements on matters of mitigation, each locality has to create and develop its own, especially since many functions performed formerly by the national government, including those relating to agriculture, environment and human welfare, have been devolved to the local government units (LGUs) by virtue of the Republic Act 7160 of 1991. Devolution enables local autonomy, allowing LGUs to decide which policies to adopt and which action programmes to finance.

The main objective of this chapter is to examine the position of Misamis Oriental (MisOr) LGU in terms of policies and actions apropos the national CCRF in the context of local autonomy, using Talcott Parsons' (1960) social systems theory. This theory hypothesises that MisOr LGU is in one accord with the national government. It argues that key actors – namely, government in polity, business in the economy and civil society organisations in society – have to achieve calibrated interdependence in order to effect substantial ground-level CC adaptation among the populace.

MisOr province has been selected for this research study for the following reasons. First, it was one of the two provinces selected by the Human Development Network (HDN) as a pilot site to examine the LGU utilisation of statistical data in planning and policy-making. Second, the fundamental criterion for selecting a province was the existence of a stable research institute in the locality to be a partner of the HDN for the study. Since the Research Institute for Mindanao Culture established in 1957, is in MisOr, the province was chosen by the HDN.

This chapter consists of five main sections, excluding the introduction and conclusion – namely, (i) a literature review and theoretical framework, (ii) the research methods, (iii) the findings, (iv) a discussion and analysis and (v) limitations. The CCRF of the Philippines is discussed first. A theoretical framework for analysis is also introduced. This chapter also examines various environmental issues and challenges faced by the MisOr province. From the findings and discussion, both negative and positive lessons are derived which could be useful for the implementation of the CCRF.

2. Literature review and theoretical framework

2.1 Climate-change response framework

The Presidential Task Force on Climate Change (2007) named four imperatives for CC mitigation and adaptation – namely, (i) the country's participation in international action, (ii) the effective governance of the CC response, (iii) proaction and cooperation among diverse stakeholders and (iv) lifestyle changes. These imperatives called for a climate-friendly energy supply mix, policy incentives for renewable energy, diverse interventions in energy generation, and adaptation to address issues with particular sectors and specific geographic areas. Other initiatives include focusing on high-risk settlements, as well as the vulnerable population centres and food-producing areas, financing interventions in support of local and sectoral initiatives for market-based incentives targeted at lowering the costs of power generation using clean technologies, technology solutions for low-carbon energy infrastructure in agriculture, industry, transportation and settlements, and social mobilisation of various stakeholders (e.g. home and office builders, local government officials and car manufacturers). These imperatives formed an interdependent structure that corresponds with the theoretical operations of structural functionalism, which theorises that society is a complex system, and various parts of the system must work closely with each other in order to make it function properly. This posits a multiway interaction among systems interchanging with other systems in the environment while maintaining an intrasystem equilibrium (Parsons, 2007).

Parsons (1960) formulated a paradigm called adaptation, goal attainment, integration and latency (AGIL: an adaptive system representing the economy; goal attainment representing polity; an integrating system of norms or a legal system; and a latency or value system, achieved in socialisation or pattern maintenance in culture). Parsons (1960) explained that adaptation reflects the capacity of the living system to cope actively with the environment.

To function properly, each system gets inputs from other systems. The economic system, for example, requires a contribution from polity (i.e. state's support for businesses). The correspondence between the CCRF and Parsons' AGIL is shown in Table 15.1.

This study uses the framework of functional role interdependence among key actors – namely, business in economy, government in polity (MisOr LGU), and civil society organisations in society (ASEAN Regional Centre of Biodiversity Conservation, 2002; Ha, 2012a, 2012b). Parsons (1960) referred to functional interdependence as role interaction such that in physical space, motivated by the drive towards the 'optimisation of gratification', actors interact with one another in a manner defined by a system of culturally shared symbols. The tendency towards the 'optimisation of gratification' implies that actors participate in situations that bring back benefit, and therefore the goals of environmental governance at the local level are better achieved when people's engagements in mitigation and adaptation offer them positive returns.

Business deals primarily with production/distribution of goods/services, which must be suitable for consumption. Government promotes democratic governance for goal attainment with transparency, accountability and respect for human rights. Civil society provides important motivators for lifestyle modification such that people's behaviour changes towards refocused worldviews, values and aspirations, and it moves in the direction of the more sustainable consumption behaviour, one that goes past the expectations of public policy (OECD, 2008).

Government regulates the establishment of business. Businesses influence government and the civilian community by the goods and services

Table 15.1 Theoretical framework: Climate change adaptation framework and Parsons' AGIL

Elements in Philippine climate-change adaptation framework	A Adaptation (economy)	G Goal attainment (polity)	I Integration (legal system)	L Latency (enculturation)
governance and climate-change response	governance of economic adaptation	goals set and pursued by government	legal bases of climate change response	governance of climate-change response norms embedded in culture
stakeholder involvement	business adaptation	conformity of business activities with polity goals	adherence of business to legal provisions	conformity of business ethics with cultural values
lifestyle change	adaptation in consumption patterns	collaboration between civil society and government in polity	lifestyle change in conformity with ethical-legal expectations	climate-change adaptive consumption patterns promoted by civil society
international commitment	economic adaptation vis-à-vis globalisation	international commitments attuned to polity goals	legal provisions attuned to international agreements	continuities between cultural practices and international commitments

Source: the first author.

they produce and how they are produced. The economic opportunities, products and services that businesses provide shape, and are shaped by, people's culture and behaviour. Government provides legislation and services to both businesses and people in the community. Civil society influences business and government through pressure groups, advocacies and the media (Teegen, Doh and Vachani, 2004).

2.2   Climate-change perspectives

The literature contains dissonant tones, just as different Philippine provinces respond to CC differently. While the World Business Council for Sustainable Development observed that the consumption behaviour of consumers worldwide puts undue stress upon the earth's ecosystem, some economists saw antithesis in the long term, as evidenced by the environmental Kuznets curve studies refuting claims that environmental degradation was an inevitable consequence of economic growth (Stiglitz, 2005). This dissonance afforded decision-makers some options – for example, some stuck to one side or the other while others escaped between the horns.

Canada, tagged in one report as among the earth's worst polluters (together with Australia and the United States), expressed general reluctance to reduce its greenhouse gas emissions. The Philippines, an economic slowcoach, was the lowest among the countries in Southeast Asia in per capita CO_2 emissions from energy use but was among the first to sign the UNFCCC in 2001 (Goco, n.d.). Since CC pertains not only to physics and chemistry but also to economics and politics, *inter alia*, the Kyoto Protocol allowed the trading of pollution credits. A country can buy pollution credits from another country where reducing emissions is cheaper, and have these purchased credits registered as part of their own quota (Secretariat of the UNFCCC, 2012).

On the whole, consensus tended to build towards preferential option for sustainable human progress through environmentally sound transformations in business, government and civil society, each according to its own capacity, function and accountability (Hysing, 2009; Drexhage and Murphy, 2010; Avelino, 2011; Fisher and Surminski, 2012; Falaschetti, 2013).

3. Research methods

Data for this chapter were taken from both academic and non-academic sources. Non-academic sources include records and service statistics of the MisOr provincial government, and the regional and provincial offices of the national government line agencies, from published sources and from the responses of planning officers in municipalities to questionnaires in a survey.

The authors organised 12 knowledgeable raters to rank the municipalities comprising MisOr according to business and industry concentration. Their rankings yielded a Kendall Coefficient of Concordance, W, of 0.72, significant at the 0.01 level or better with a chi square test (Legendre, 2005). Triangulation with data from the Department of Trade and Industry (DTI), a government agency tasked to issue registration certificates to establishments, confirmed the rankings. There were 6 municipalities with 'high' business and industry concentration, 11 with 'moderate' and 8 with 'low'.

4. Research findings

4.1 The context

The Republic of the Philippines has a central government. It is divided into provinces run by LGUs. The province is divided further into municipalities and component cities, which in turn are composed of *barangays* (roughly, 'villages').

Located in northern Mindanao, MisOr is composed of 25 municipalities including two component cities. It has a total land area of 298,170 sq. km, and a total population of 103,775,002 (July 2012 estimate) (CIA World Factbook, 2012). It experiences rain practically throughout the year, some of which comes with typhoons. Its main crops are rice and corn, but it also exports, among others, crude coco oil, canned pineapple products, fatty alcohol, sintered ore, desiccated coconut, copra expeller, cakemeal pellets, milk powder, activated carbon, finished wood products and cold-rolled-steel sheets (Provincial Government of Misamis Oriental, 2011).

Most of the local environmental legislation and policy enactments in MisOr were made in response to challenges to the environment posed by large industries and business establishments. Type A municipalities have at least 27 large, mostly multinational, establishments. Type B municipalities have poultry and hog farms, cottage industries and some factories. Type C municipalities have few of these or none at all (Provincial Government of Misamis Oriental, 2011). Air and water pollution were the most common issues that the interview respondents identified in Type A municipalities; water pollution and soil erosion were most common in Type B municipalities; and soil erosion was most prevalent among Type C municipalities. These problems were not mutually exclusive, of course. Yet some were more obvious in certain municipality types than in others.

Mining is conspicuous by its absence in the list of industries in the municipalities. In 2009, the country's gross production value from large-scale metallic mining, small-scale gold mining and non-metallic mining was PhP82.1 billion. The gross value added was PhP96.9 billion and the mining contribution to the gross domestic product was 1.3 per cent (Mines and Geosciences Bureau, 2010). Some 57 large-scale mining operations were in Mindanao (Quitoriano, 2009). Despite these gains, there was no clear evidence of the benefits that mining brought to the people, especially in the 'sites of struggle' (SOS is the term used to refer to communities in the mining areas, or areas with extractive operations) after more than 15 years of the implementation of the Mining Act 1995 (Sealza, 2010).

Many foreign firms were reported to get around the law requiring certain standards for environmental protection and compensate by using Filipino-owned small-scale mining cooperatives (Bergonia, 2011). The law on small-scale mining had lower environmental standards. Furthermore, these firms were reported to bribe LGUs in order to gain entry into mining sites. 'Everyone can be paid,' was how local residents summarised corruption as they expressed hopelessness in opposing the entry of mining operations approved by the LGU (Junio, 2008; Macabuac-Ferolin, 2008).

4.2 Environmental issues and challenges

Unforeseen inundation and drought are two of the more immediate effects of CC that trigger a series of other effects. In 2009, MisOr was hit by a strong typhoon that caused flooding and the displacement of some 20,000 people (Baños, 2009). The following year the *El Niño* phenomenon resulted in double-digit declines in coconut and sugarcane production (Jegillos, 2000). In December 2011, Tropical Storm *Sendong* (international codename *Washi*) hit northern Mindanao. More than 2,000 people were reported dead and hundreds missing after the storm (Office of Civil Defence Regional Disaster Risk Reduction and Management Council – Region X, 2012). Many of the dead bodies were said to have been found under clean-cut timber believed to have been carried down by strong water current from illegal log ponds upstream.

Air- and water-quality issues raised in Type A municipalities were confirmed by two independent studies at the College of Medicine at Xavier University, one on air and the other on water. The particulate matter (PM_{10}) content of air in the area (27.3 $\mu g/m^3$) exceeded the upper tolerable limit of 20.0 $\mu g/m^3$ (for PM_{10}) set by the World Health Organisation (WHO, 2006; Jamis *et al.*, 2008). The second study showed that the extent of faecal contamination in coastal areas of Misor far exceeded the standard for clean sea coasts. The Most Probable Number index of 200/100 ml of water has been set by the Department of the Environment and Natural Resources (DENR) as the tolerable faecal coliform level. However, the municipalities along the MisOr coastline had indices ranging from 220/100 ml to 1600/100 ml (Taruc *et al.*, 2008).

Environmental studies at McKeough Marine Research Centre, Xavier University, showed that

- coral reef cover along the coast of MisOr had declined from 59 per cent in 1997 to 38 per cent in 2008;
- the average biomass of reef fishes in MisOr bay was in the lower-medium category;
- coastal waters of Cagayan de Oro City had a low biomass (4.0 metric tons/km^2);
- municipal fisheries production indicated non-sustainable conditions having declined by about 25 per cent over the past decade (Roa-Quiaoit *et al.*, 2010).

4.3 Responses of systems

4.3.1 *Government in polity*

Given sufficient environmental policy and implementation, industry and trade could be environmentally friendly (Medalla, 2001). Hence it is crucial that environmentally sound policies and the political will to carry them out are firmly in place. In order to execute properly the environmental provisions for national government agencies, for example, the DENR-Department of the Interior and Local Government Joint Memorandum Circular

No. 98-01; Philippine Agenda 21, the executive branch of the LGU, established the Provincial Environment and Natural Resources Office (PENRO) in August 2005. This represented a major institutional structural change because it meant funding, human resources allocation, procedural adjustments and so on. Previously, environmental concerns were adjunct to the Provincial Agriculture Office.

Regarding policy, in 2006 the legislative branch of MisOr provincial LGU passed a comprehensive Environmental Code for use by the PENRO. This The Code contained, among other provisions, the following:

- protecting the atmosphere through air- and noise-pollution management, ecological and ecosolid-waste management, and air-quality management;
- proaction and cooperation among stakeholders by gathering data to clarify key issues and concerns, and to characterise the environmental setting, predict the impact, and measure the social acceptability of the project; and, creating an environmental monitoring team headed by the provincial governor with members from the LGU, the private sector and government line agencies;
- arresting deforestation by providing fund allocations for forest-resources management in the municipal and provincial budgets, and regulating mining and quarrying;
- conserving biodiversity through ecotourism (see the environmental code of MisOr).

The code regulated business operations, created structures for environmental monitoring, allocated budget for forest-resources management, provided procedures to protect water resources, and empowered municipal LGUs to manage, rehabilitate and maintain small watershed areas. Hence, on the level of action, some activities in the province were:

- protecting the environment by organising the Provincial Solid Waste Management Board with a multisectoral composition, and supporting the Protected Area Management Board initiatives by constructing sanitation facilities, such as the pay toilets;
- protecting the supply and quality of freshwater by laying down the groundwork for the PhP30M 'Project for Enhancement of Barangay Governance and Community Empowerment in Micro-watersheds in Misamis Oriental' under the technical cooperation programme of the Japan International Cooperation Agency;
- planning and managing land resources using the integrated approach, establishing the Misamis Oriental Landcare Initiative, formalising cooperation among the provincial government and concerned national

government agencies, the World Agroforestry Centre and the Landcare Foundation, Inc. to jointly advocate for intensified natural resources management;

*   protecting the ocean and coastal regions by rehabilitating mangrove and forest areas. (Balay Mindanaw Foundation Inc., 2008).

As shown in the office records available upon request, the province also continued to increase its environmental budget allocations as a percentage of the total budget (e.g. 1.2 per cent in 2003; 6.7 per cent in 2004; and 9.2 per cent in 2005) (Provincial Planning and Development Office, 2011) The individual municipalities also responded. In 2003, 18 out of 23 municipalities and two component cities provided budget for environmental projects at an average of 1.1 per cent of the total municipal budget. In 2004 and 2005, 20 municipalities made such provision at an average of 1.7 per cent and 1.6 per cent, respectively. Finally, houses in disaster-prone areas were relocated (Provincial Planning and Development Office, 2011).

4.3.2   Business in the economy

Municipalities with a high level of industry concentration (Level A) had consistently earned more – three to four times as much – than the other types of municipalities over the three-year period (2003–2005). In the Philippine ranking scheme, Type A municipalities were, on average, 3rd Class in 2005. Type B municipalities were 4th Class and Type C municipalities were 5th Class (Provincial Planning and Development Office, 2011).

The responsibility over threats to life is greater in municipalities with a higher concentration of industrial plants. Trade-offs to justify a community's tolerance of environmental hazards – for example, financing services to manage the negative effects of industries on health and livelihood – were expected from industries. Businesses in Type A municipalities, however, had limited provisions that directly enabled community residents' adaptation to changes occasioned by industrial operations. In addition, industries generally appeared to have no environmental protection programmes outside government requirements for a licence to operate. The surrounding communities seemed to only have indirect benefits from business and industry in some human welfare indices. For example, compared with other types, Type A municipalities had the highest child immunisation rate and the highest scores in educational performance indicators, on average.

Issues of air, soil and water pollution had implications upon which agency issued the business permits, collected payments and regulated/monitored operations. Agri-based industries, such as livestock, had to register with the DTI. These types of industry also had to get permits to operate from, and therefore had to be regulated by the LGU. Hence they had to pay taxes locally. Many large industries, however, had the option not to register with the local DTI. They were granted permits by special legislation and were being regulated by a national administrative structure, such as the DENR or the Philippine Economic Zone Authority. Thus the LGU had very little control over these industries.

4.3.3   Civil society organisations in society

Many civil society groups, some of which had support from international environmental non-government organisations (NGOs), had been involved in the ridge-river-reef projects. A case in point was the Cagayan de Oro River Basin Council formed from a Climate Change Congress in 2010. The group had conducted water tests and information campaign activities among residents along watersheds and coastlines (Jorgenson, Dick and Shandra, 2011).

Likewise, a group led by an association of volunteer lawyers once required that a candidate, in order to get their members' votes, should have a 'green platform' – that is, the candidate had to commit to the following: enactment of the water code; implementation of waste segregation and a comprehensive drainage system; support of organic and diversified farming; relocation of communities from hazard-prone areas; implementation of clean air and the anti-smoke-belching ordinance; and helping to contain global warming (Sun Star Publishing Inc., 2010; Mix, 2011). Many politicians ran with the 'green platform' and won in the 2010 elections.

Couples for Christ-*Gawad Kalinga*, a church-based lay organisation, has been helping flood victims and other vulnerable groups to build homes as they try to adapt to CC, in some cases in cooperation with the LGU. In addition, other people have formed volunteer rescue groups for flood contingencies.

5. Discussion and analysis

Notwithstanding local autonomy, MisOr LGU endeavours were in accord with the national CCRF and therefore, with international CC agreements, insofar as the province's goal attainment and integrative functions were concerned. However, it had limited control over operations of industrial giants inside special economic enclaves created by the national government, over dubious mining and logging businesses that emerged through corruption and loopholes in the law, and over unregistered small enterprises in the informal economic sector. In its desire to attract and eventually retain foreign investments, government apparently granted exemptions to industrial giants regarding environmental regulations for fear of capital flight.

Similarly, while the LGU had legal measures to regulate household economic activities and waste disposal, none of these were rigidly carried out. Civil society took minute steps to change culturally established livelihood and waste-disposal practices. Its priority mainly centred on the more controversial sources of environmental destruction. Business establishments did not have many CC adaptation schemes to help the residents who were affected by their operations.

Parsons' functionalist social systems theory called for business in economic adaptation to respond to the goal-attainment and legal integration functions of government in polity, as business needed protection from government. By the same token, civil society was expected to manifest cultural latency vis-à-vis business and governance because adaptive efforts, goals and rules could only be as good as the citizens' commitment to abide by them. Government, civil society and business subsystems fell short of the kind of interdependence expected in structural functionalism. So far, conscious cooperation to address CC concerns among business in the economy, government in polity and civil society in culture was inadequate. LGU's orchestration effort was required. Proof of inadequacy can be seen when government, business and civil society were helpless when tropical storm *Washi* struck. Although conscious effort is not the only major requirement in functional interdependence, equilibrium is better achieved among systems when each one is conscious of being able to effectively demand from, and serve the needs of, the others in the system.

It is recommended that the LGU and the national government put their acts together in the matter of regulating business. For example, transnational investors should be required to bring in only the environmentally sound technologies for productive or extractive activities. It is also recommended that since the LGU was mindful of the environmental threats due to big industrial plants, it should be given a hand in instituting mitigation and adaptation requirements for them.

Illegal mining and logging should stop. Laws on small-scale mining have to be revisited. With the large number of small-time miners operating in an area, there is no such thing as small-scale mining. Also, legal logging concessions bring into the same log pond logs that have been cut from outside the concession. In this regard, civil society should police its own ranks because politician-organised businessmen-organised NGOs can muddle legitimate issues for political or business gains, both in goal attainment and integrative functions of government, to the detriment of efforts in CC mitigation and adaptation.

Existing rules, orders and ordinances that pertain to the environment may have to be re-examined for their adequacy of positive and negative reinforcements. Are they potent enough to elicit obedience? What perceived and factual, tangible and intangible gains will people benefit from by observing them (e.g. what does the household get from waste segregation)? What penalties are provided for and meted out against violators? Are people aware of the incentives and disincentives? What are the prevailing attitudes and practices?

In many cases, small- and medium-sized enterprises (SMEs) and industrial enterprises, registered or unregistered, are allowed for as long as their
wastes levels are within tolerable limits. The acts of goodwill that these enterprises extend to the community are usually enough to make their presence
acceptable. The LGU should organise a team comprising various members
from government, business and civil society to monitor these enterprises
(Ha, 2012a, 2012b).

The environmental pollutants that emanate from households need to be
quantified as well so that decision-makers can set tolerance limits. Human-
istic considerations temper the enforcement of rules. Many SMEs lack the
capability to meet the environmental standards set by the authorities.
However, for many poor households, these SMEs are their only source
of livelihood, hence they are being left alone. In the end, the sense of
responsibility, and therefore latency, is weakened, and the monitoring and
evaluation activities are ignored.

6. Limitations

The main limitation of this chapter is that since MisOr was chosen purposively, the environmental legislation and the position that the province
was taking vis-à-vis the national CCRF does not reflect the legislation and
position of other Philippine provinces. Also, the findings may be pertinent
only to MisOr, and the many factors that affect the implementation and
outcomes of environmental activities may vary from province to province.

7. Conclusion

This chapter has discussed the national CCRF and how local governance
agencies have responded to the framework in terms of policies and actions
in the context of local autonomy, using MisOr as a case study. Talcott
Parsons' (1960) social systems theory has been adopted to explain why interaction among systems in the environment is important in environmental
governance.

Having experienced losing lives from inundation, MisOr supported international accord through legislation – for example, land and water resources
management to combat deforestation and conserve biodiversity. It has created an environmental institute, invested in human resources, increased
environmental budget, created a waste-management body, and laid groundwork for a watershed and land-care coalition with civil society organisations.
However, operationally, the local government as one of the key actors in
the system has exhibited subdued political will, and civil society organisations have struggled to get their acts together. Business as the key actor in
the economy failed to regulate itself and continued to pose challenges to

environmental governance. Civil society fell short of expectations while it has started to cooperate in rehabilitating ridge, river and reef. Thus it is recommended that various groups of actors have to work closely with each other in the governance system to mitigate and adapt CC. If one part of the societal system does not function appropriately, other parts will be weakened.

Future directions of research should focus on the interaction of government, business and civil society in polity, economy and culture as a whole domain.

Abbreviations

AGIL	adaptation, goal attainment, integration and latency
CCRF	Climate Change Response Framework
CC	climate change
DENR	Department of Environment and Natural Resources
DTI	Department of Trade and Industry
HDN	Human Development Network
LGUs	local government units
MisOr	Misamis Oriental
PENRO	Provincial Environment and Natural Resources Office
PM	particulate matter
SMEs	small and medium-sized enterprises
SOS	sites of struggle
UNFCCC	UN Framework Convention on Climate Change
WHO	World Health Organisation

References

ASEAN Regional Centre of Biodiversity Conservation (2002) 'Fulfilling the Goals of Agenda 21', *Asean Biodiversity*, July–September 2002, 36–41.

F. Avelino (2011) Third Sector Research and the Power of Civil Society in Sustainability Transitions. Paper presented at the 2nd International Conference on Sustainability Transitions Diversity, Plurality and Change: Breaking New Grounds in Sustainability Transition Research, 13–15 June 2011, Lund, Sweden: Lund University.

Balay Mindanaw Foundation, Inc. (2008) *Project of Enhancement of Local Governance and Community Empowerment in Micro-watersheds in Misamis* (Misamis Oriental, Philippines: Japan International Cooperation Agency and Provincial Government of Misamis Oriental).

M. Baños (2009) Flass Floods Displace 20,000 in Misamis Oriental, www.cbcpnews .com/?q=node/6828, date accessed 26 August 2011.

T.S. Bergonia (2011) 'Chinese Firms Skirt PH Laws', *Philippine Daily Inquirer*, May 21, 2011.

CIA World Factbook (2012) Philippines. CIA World Factbook, https://www.cia.gov/ library/publications/the-world-factbook/geos/rp.html, date accessed 19 February 2013.

J. Drexhage and D. Murphy (2010) *Sustainable Development: From Brundtland to Rio 2012* (New York: United Nations).

D. Falaschetti (2013) *Global Environmental Governance: Mechanism Design Lessons from Corporate Governance* (Bozeman MT: The Property and Environment Research Centre).

S. Fisher and S. Surminski (2012) *The Roles of Public and Private Actors in the Governance of Adaptation: The Case of Agricultural Insurance in India* (UK: Centre for Climate Change Economics and Policy, Munich Re Programme Technical, and Grantham Research Institute on Climate Change and the Environment).

J.A. Goco (n.d.) *The UNFCCC, Kyoto Protocol and an Introduction to CDM* (Quezon City Philippines: IACCC Secretariat).

H. Ha (2012a) 'A Multi-sector Governance Model for Environmental Sustainability – Australia Case', in J.R. Barker and R. Walters (eds) *New Zealand and Australia in Focus: Economics, the Environment and Issues in Health Care* (USA: Nova Science Publishers, Inc.), pp. 35–60.

H. Ha (2012b) Networked Governance Model for Sustainable Development. Paper presented at The 2nd International Conference on International Relations and Development (ICIRD 2012), Towards an ASEAN Economic Community (AEC): Prospects, Challenges and Paradoxes in Development, Governance and Human Security, 26–27 July 2012, Chiang Mai, Thailand.

E. Hysing (2009) 'From Government to Governance? A Comparison of Environmental Governing in Swedish Forestry and Transport', *Governance: An International Journal of Policy, Administration, and Institutions*, 22, 4, 647–672.

L.V. Jamis, A.B. Aliman, S.J. Lacaden, R.A. Laplap, A. Mascarinas, E. Obouyes, S.Z. Sumndad and H.L. Tan (2008) *An Analytical Study of the Relationship Between Selected Respiratory Diseases Among Adult Patients Admitted at the Northern Mindanao Medical Centre and the Levels of Air Pollutants in Cagayan De Oro City, 2005–2007, Student Working Series No. 16* (Cagayan de Oro City: KRC, Xavier University).

S.R. Jegillos (2000) *Reducing the Impacts of Environmental Emergencies: The Case of the 1997–1998 El Niño Southern Oscillation. The Philippine Study* (Manila: Asia Pacific Disaster Management Centre, Inc.).

A.K. Jorgenson, C. Dick and J.M. Shandra (2011) 'World Economy, World Society, and Environmental Harms in Less Developed Countries', *Sociological Inquiry*, 81, 1, 53–87.

R. Junio (2008) Balabag Exploration Project: Understanding Stakeholders and Analysing Conflict. Paper presented at the Conference on Mining in Mindanao, February 28, Cagayan de Oro City: Xavier University.

P. Legendre (2005) 'Species Associations: The Kendall Coefficient of Concordance Revisited', *Journal of Agricultural, Biological, and Environmental Statistics*, 10, 2, 226–245.

K. Lofts and A. Kenny (2012) *Mainstreaming Climate Resilience into Government: The Philippines' Climate Change Act* (London: Climate and Development Knowledge Network).

M.C. Macabuac-Ferolin (2008) Kabkab Hako:! Níkel Mining in Claver, Zurriago del Norte. Paper presented at the conference on Mining in Mindanao, February 28, Cagayan de Oro City: Xavier University.

M.M. Medalla (2001) 'Environmental Impact of Trade Policy Reforms on Pollution Intensity', *Philippine Journal of Development*, 28, 2, 167–204.

Mines and Geosciences Bureau (2010) The Philippines Minerals Industry at a Glance. http://www.mgb.gov.ph/, date accessed 26 March 2010.

T.L. Mix (2011) 'Rally the People: Building Local-Environmental Justice Grassroots Coalitions and Enhancing Social Capital', *Sociological Inquiry*, 81, 2, 174–194.

OECD (2008) *Promoting Sustainable Consumption: Good Practices in OECD Countries* (Paris: OECD).

Office of Civil Defence Regional Disaster Risk Reduction and Management Council – Region X (2012) *Tropical Storm Sendong: Post Disaster Needs Assessment* (Quezon City: Office of Civil Defence Regional Disaster Risk Reduction and Management Council – Region X).

T. Parsons (1960) 'A Sociological Approach to the Theory of Organisations', in O. Grusky and G.A. Miller (eds) *The Sociology of Organisations Basic Studies 1970* (New York: The Free Press Macmillan).

T. Parsons (2007) 'An Outline of the Social System', in C. Calhoun, J. Gerteis, J. Moody, S. Pfaff, and I. Virk (eds) *Classical Sociological Theory* (2nd edition) (Malden, MA: Blackwell), pp. 421–440.

Presidential Task Force on Climate Change (2007) Climate Change Response Framework, http://www.doe.gov.ph/cc/ccrf.htm, date accessed 27 August 2011.

Provincial Government of Misamis Oriental (2011) Geography, http://www.misamisoriental.gov.ph/index.php/profile/geography, date accessed 15 September 2011.

Provincial Planning and Development Office (2011) Misamis Oriental Provincial Development Plan. Provincial office document.

E.L. Quitoriano (2009) *Land, Foreign Aid and the Rural Poor in Mindanao* (Bangkok: Belgian Alliance of North-South Movements and the Focus on the Global South-Philippines).

H.A. Roa-Quiaoit, A.B. De Guzman, E.A. Villaluz, D.R. Dawang, F.T.S. Quimpo, A.S. Mabao and L.S. Martinez (2010) *Ecological and Fisheries Profile of Macajalar Bay* (Cagayan de Oro City: MMRC, Xavier University Press).

I.S. Sealza (2010) Mining and Ancestral Domain: Some Issues in Mindanao. Paper presented at the 2010 National Conference of the Philippine Sociological Society, 14 October 2010, Dumaguete City: Silliman University.

Secretariat of the UNFCCC (2012) *International Emissions Trading* (Bonn, Germany: Secretariat of United Nations Framework Convention on Climate Change).

Sun Star Publishing Inc. (2010) Group Pushes 'green agenda'. http://www.sunstar.com.ph/cagayan-de-oro/group-pushes-%e2%80%98green-agenda%e2%80%99, date accessed 19 February 2013.

J.E. Stiglitz (2005) 'The Ethical Economist: Growth May Be Everything, but It's Not the Only Thing', *Foreign Affairs*, 84, 6, 128–134.

L. Taruc, E. Butron, J. Casan, R. Gumban, C. Lagunilla, C.M. Rosales, O.G. Spalleda, M. Tadifa and A.R. Uba (2008) *A Descriptive Study of the Extent of Faecal Contamination and the Factors Affecting Contamination of Macajalar Bay*, Student Working Series No. 17 (Cagayan de Oro City: KRC, Xavier University).

H. Teegen, J.P. Doh and S. Vachani (2004) 'The Importance of Nongovernmental Organisations (NGOs) in Global Governance and Value Creation: An International Business Research Agenda', *Journal of International Business Studies*, 35, 463–483.

WHO (2006) *WHO Air Quality Guidelines for Particulate Matter, Ozone, Nitrogen Dioxide and Sulphur Dioxide: Global Update 2005* (Geneva: WHO).

16
Responses to Climate Change – Who Is Responsible? A Conclusion

Tek Nath Dhakal and Huong Ha

1. Background

The impacts of climate change (CC) are real and palpable, posing threats on several fronts, including human development and the sustainability of civilisations. It has been agreed by all authors that CC impacts are profound, long-lasting, cross-border and difficult to contain at a location or within a country. The major impacts of CC identified in this book include (i) loss of life, human security and human displacement, (ii) ecosystem degradation and biodiversity loss, (iii) economic loss and reduction of economic growth rate and (iv) national and regional security. These have been witnessed as there have been sea-level rises, glacial retreats, adverse weather patterns, flooding, shrinking freshwater resources, desertification and loss of bio-diversity at an alarming rate. The International Centre for Integrated Mountain Development (ICIMOD) (2011) study has shown that the Hindkush-Himalaya region is a hotspot of global warming, recording a temperature rise of 0.6–1.3 °C within a period of 30 years. CC and its adverse effects are pushing humans to a point where we are living in a precarious condition (Regmi and Dinanath, 2011). Thus, as Huq and Reid (2006) also identified, CC is now considered to be one of the major environmental problems challenging society and the natural world. There has been no major difference in terms of the impacts of CC in Asian countries and others. Yet Asian countries are prone to disaster due to their geographical, social and cultural uniqueness, which makes the battle against CC and environmental degradation even more challenging.

The state of environmental governance by different sectors has been discussed in this book. It also focuses on the role of the different actors in governance and practical steps that need to be taken on adaptation and mitigation, water, pollution, droughts, floods, natural disasters and so on, pertinent in different region and countries. The impacts of CC are widespread and transnational, and thus it requires the contributions of different sectors of society and different groups of stakeholders within each region. National policy initiatives, macro- and micropolicies involving social and economic measures and state-led governance innovations, which are supported by initiatives from other non-state sectors (e.g. the private sector and civil society), to mitigate the adverse effects of CC on national development plans and outcomes, has also been examined in this project.

2. Concerns of climate change

The global mechanisms for dealing with CC issues, such as the United Nations Framework Convention on Climate Change (UNFCCC), and the Intergovernmental Panel on Climate Change (IPCC) under which global negotiations are held, must be enabled to work efficiently and effectively to tackle the growing menace of CC in an equitable and just manner. The recent Durban Climate conference has been able to make some headway in the direction of ensuring operationalisation of the Green Climate Fund in favour of the developing countries, including least developed countries and globally binding emissions being cut by 2020. The Kyoto Protocol has been extended beyond its expiry in 2012. In this context, Rio+20 at the Earth Summit in 1992 provided a unique opportunity to agree on, and commit to, making the 'change-of-course' and for broad acceptance of fundamental changes that are needed to manage the activities through which human actions impact on the earth's sustainability. The implementation of such commitments is obviously challenging because there is competition and conflict over scarce resources. One of the best ways out would require a degree of cooperation globally, regionally and locally to solve the problems caused by CC. Given the urgency of the problem, the way in which these mechanisms are working needs to be streamlined and made more effective to ensure more concrete results on the ground. The World Bank (http://www.forestcarbonpartnership.org) has set up the Forest Carbon Partnership Facility (2013) with two major objectives: (i) building capacity for formulating policy, plans and programmes relating to REDD+ (enhancing carbon storage in forest ecosystems) through its technical assistance programme, and (ii) testing REDD+ projects with performance-based incentive payments in some pilot countries in order to set the stage for a much larger system of positive financial incentives through voluntary and market mechanisms in the future (Upadhyay and Bhattarai, 2011).

The responsibility for safeguarding CC not only remains within the international framework; concerns about CC at the regional, national and local levels are equally important. Human needs, the pace of development and the severity of CC are found in different magnitudes. Regional organisations such as the South Asian Association for Regional Cooperation and the Association of Southeast Asian Nations, should work in solidarity and with a sense of purpose to fight the menace of CC in a comprehensive manner, exerting moral pressure on the developed world and relevant international organisations for deeper and speedier emissions cuts and an increased flow of resources, technology and funds for poor countries to cope with their growing need to adopt appropriate mitigation and adaptation measures.

3. Who is responsible?

Though the CC phenomenon is intriguing and might have been the result of several contributory factors, the contribution of man-made industrial and vehicular emissions is key. In this sense the industrialised countries and those developing countries with high growth trajectories are responsible for increased emissions and hence their contribution to global warming and CC is greatest. However, countries like Nepal whose contribution to greenhouse gases is minimal is equally vulnerable to CC as the negative effects defy national borders. Even within national borders, the poor and the marginalised sections of society are found to be most vulnerable to the impacts of CC. Knowingly or unknowingly the poorest sections of society are also putting pressure on ecosystems through deforestation for the sake of their livelihoods. As a result, these poorest of the poor who live closer to nature have both their lives and their livelihoods at stake. It is therefore a moral obligation for the world community, especially the economically advanced countries, to ensure climate justice for the poorer parts of the world and the poorer sections of humanity.

The equity and justice aspects of CC are very important for developing countries such as Nepal. The most CC-vulnerable people live in these countries and they are paying the price for no mistakes of their own. This situation calls for improvement of natural environments through mechanisms like the Clean Development Mechanism, the Kyoto Protocol, reducing emissions from deforestation and forest degradation (REDD) and REDD+, ameliorating the living conditions and livelihoods of the poorest of the poor, providing alternative and renewable sources of energy for them, improving agricultural practices, conserving water resources, and ensuring other appropriate mitigation and adaptation measures to ensure a shared progress and prosperity.

Yet, several problems surface during this course of action. For example, CC-related policies and legislation are available but there has been disagreement about the approaches and mechanisms of responding to CC among the parties involved. Cooperation among sectors is fragile due to a lack of a common understanding about the priorities of issues, a lack of capacity and capabilities, and insufficient financial and physical resources.

Several models of governance in various locations within a country and in different Asian countries have been explored. However, there is no best governance approach which can apply 'wholesale' to all locations, all sectors and all countries. Also, the current governance approaches to mitigate and adapt CC impacts are either ineffective or outdated. Thus many authors have proposed various modified versions of the current governance approaches, embracing the public and private sectors, and civil society, to prevent and reduce the adverse effects of CC effectively and efficiently.

Thus tackling of the multifarious problems emanating from CC, only a multistakeholder approach that also discusses the three-sector governance model may have significant roles. This model involves government, the private sector and civil society organisations working together and would be able to produce the desired results. Strengthening political, economic and social governance at the global, regional, national and local levels is therefore crucial to the success of any mitigation or adaptation measures that the world community has been able to design or innovate so far. It must ensure a flow of resources, know-how and technology from the richer regions to the poorer ones on the one hand, while enhancing the capability of the receiving states to shape their own destiny on the other. As stated by Ha (2011, p. 11), the capacity to produce the effect of the three-sector governance model depends on several factors, such as 'leadership commitment and competency, public trust, clear strategic plans, sufficient fund, effective enforcement of laws, public education'. In addition, 'efficiency and effectiveness of institutions, clear roles of relevant agencies, continuous monitoring and evaluation of the implementation of CC policies' are important for the success of climate governance (Ha, 2011, p. 11). Due to the lack of a strategic roadmap and architecture on climate financing and governance, the developing countries, such as Nepal, lack clarity, and there is a dilemma around the institutional and financial governance structure to drive the agenda forward. The other issues are capacity and human resources, information and knowledge gaps, and severely lacking serious discussion regarding climate governance.

4. Conclusion

To attract the attention of the concerned stakeholders, there have been ongoing debates at various levels about CC, and the importance of streamlining the necessary policies and implementing plans for mitigation and adaptation of CC. The chapters in this book are the products of the Network of Asia and Pacific Schools and Institutes of Public Administration and Governance (NAPSIPAG) conference, co-organised by the Central Department of Public Administration and the Central Department of. Environmental Science, Tribhuvan University (Nepal), which was held in Kathmandu in December 2011. The theme was 'Reinventing Governance for Managing Climate Change and its Adaptation', within which nine subthemes highlighted the mitigation and adaptation of CC. The concerns showed at various levels – whether at the Earth Summit or at other regional and national level programmes – are the pronouncement cf such commitments. In addition, the intensity of negotiation at the international conference and at subsequent meetings is already putting pressure on national government and stakeholders to be more strategic and responsive to the needs of millions of vulnerable households waiting for a response and support. One of the triggering issues regarding CC is the lack of efficient climate governance, which needs due attention at various levels. Climate governance should address the impact of climate variability and change, climate policy responses, and associated socioeconomic development, which affects the ability of countries to achieve sustainable development goals. Finally, it requires an international consensus, taking account of views and interests of national, state and local stakeholders, on international and national legal regulations preparing a harmonised and transparent base for the responsibilities of contributing to the effects of global warming.

Abbreviations

CC	climate change
ICIMOD	International Centre for Integrated Mountain Development
IPCC	Intergovernmental Panel on Climate Change
NAPSIPAG	Network of Asia and Pacific Schools and Institutes of Public Administration and Governance
REDD	reducing emissions from deforestation and forest degradation
REDD+	enhancing carbon storage in forest ecosystems
UNFCCC	United Nations Framework Convention on Climate Change

References

Forest Carbon Partnership Facility (2013) Forest Carbon Partnership Facility (FCPF), http://www.forestcarbonpartnership.org/, date accessed 15 March 2013.

H. Ha (2011) Governance for Climate Change Management – Singapore Case. Paper presented at the 'Reinventing Governance for Managing Climate Change and its Adaptation' Conference, 15–16 December 2011 in Kathmandu, Nepal, organised by the Network of Asia Pacific Schools and Institutes of Public Administration and Governance (NAPSIPAG).

S. Huq and H. Reid (2006) *Mainstreaming Adaptation to Climate Change in Least Developed Countries (LDCs)* (Brighton, UK: Institute of Development Studies).

International Centre for Integrated Mountain Development (ICIMOD) (2011) *Glacier Lakes and Glacier Lake Outburst Floods in Nepal* (Kathmandu: ICIMOD).

B.R. Regmi and B. Dinanath (2011) Unripe Fruits or Non-Raining Clouds? Climate Change Governance and Funding Dilemma in Nepal. Paper presented at the 'Reinventing Governance for Managing Climate Change and its Adaptation' Conference, 15–16 December 2011 in Kathmandu, Nepal, organised by the Network of Asia Pacific Schools and Institutes of Public Administration and Governance (NAPSIPAG).

T.P. Upadhyay and K.R. Bhattarai (2011) Impacts of 'Reducing Emissions from Deforestation and Forest Degradation' intervention in Nepal, Paper presented at the 'Reinventing Governance for Managing Climate Change and its Adaptation' Conference, 15–16 December 2011 in Kathmandu, Nepal, organised by the Network of Asia Pacific Schools and Institutes of Public Administration and Governance (NAPSIPAG).

Index

Printed and bound in Great Britain by
CPI Group (UK) Ltd, Croydon, CR0 4YY